METHODS IN MOLECULAR BIOLOGY

Series Editor
John M. Walker
School of Life and Medical Sciences
University of Hertfordshire
Hatfield, Hertfordshire, UK

For further volumes:
http://www.springer.com/series/7651

For over 35 years, biological scientists have come to rely on the research protocols and methodologies in the critically acclaimed *Methods in Molecular Biology* series. The series was the first to introduce the step-by-step protocols approach that has become the standard in all biomedical protocol publishing. Each protocol is provided in readily-reproducible step-by-step fashion, opening with an introductory overview, a list of the materials and reagents needed to complete the experiment, and followed by a detailed procedure that is supported with a helpful notes section offering tips and tricks of the trade as well as troubleshooting advice. These hallmark features were introduced by series editor Dr. John Walker and constitute the key ingredient in each and every volume of the *Methods in Molecular Biology* series. Tested and trusted, comprehensive and reliable, all protocols from the series are indexed in PubMed.

Chondrocytes

Methods and Protocols

Edited by

Tariq M. Haqqi

Department of Anatomy and Neurobiology, Northeast Ohio Medical University, Rootstown, OH, USA

Véronique Lefebvre

Department of Surgery, The Children's Hospital of Philadelphia, Philadelphia, PA, USA

Humana Press

Editors
Tariq M. Haqqi
Department of Anatomy
and Neurobiology
Northeast Ohio Medical University
Rootstown, OH, USA

Véronique Lefebvre
Department of Surgery
The Children's Hospital of Philadelphia
Philadelphia, PA, USA

ISSN 1064-3745 ISSN 1940-6029 (electronic)
Methods in Molecular Biology
ISBN 978-1-0716-1121-0 ISBN 978-1-0716-1119-7 (eBook)
https://doi.org/10.1007/978-1-0716-1119-7

Cover Illustration Caption: The picture shows a sagittal section through the knee of a 3-month-old mice. The section was hybridized with a Sox9 RNA probe (magenta) and counterstaining with hematoxylin (blue). Sox9 is expressed in articular and growth plate chondrocytes.

This Humana imprint is published by the registered company Springer Science+Business Media, LLC part of Springer Nature.
The registered company address is: 1 New York Plaza, New York, NY 10004, U.S.A.

Preface

Chondrocyte is the only cell type present in cartilage. It fulfills essential functions for the tissue and is thereby pivotal in vertebrates in many ways. Differentiating from skeletogenic mesenchymal progenitor cells, chondrocytes build cartilage tissues at many sites in the embryo body by producing a very abundant, highly specialized extracellular matrix around them. The highlights of this matrix are a collagenous framework mainly composed of collagen type 2 and a highly hydrated gel of water, aggregates of aggrecan, and other proteoglycans and glycoproteins. On the one hand, the cartilage primordia of future long bones spatially organize into growth plates by staggering their terminal differentiation process one cell layer at a time. Growth plates ensure skeletal elongation throughout fetal and postnatal development and also participate in the progressive replacement of cartilage by bone through the process of endochondral ossification. On the other hand, the cartilage primordia that cover the opposing surfaces of bones in articular joints and that shape auditory structures and respiratory airways are maintained throughout life by chondrocytes that normally never undergo the same terminal differentiation process as in growth plates. The chondrocyte thus presents itself in multiple facets or subtypes according to the type of cartilage it resides in, its location in this cartilage, and the developmental or adult stage of the individuals. It distinguishes itself from other cell types not only by producing a unique extracellular matrix but also by exhibiting unique metabolic properties due to the avascular and aneural nature of its environment. These distinctions contribute to the fact that the chondrocyte finds itself particularly challenged in pathological conditions and is generally unable to properly respond to them to continue building or maintaining its tissue or repairing it.

Genetic, inflammatory, physical, or other alterations that affect the behavior of the chondrocyte or the cartilage matrix itself cause a large spectrum of skeletal diseases. Chondrodysplasias are cartilage developmental diseases that exist in a variety of forms, from mild dysmorphism to severe dwarfism. While each form is rare, these diseases are altogether fairly common. Hundreds of them have already been reported, and more are still being uncovered every year. Osteoarthritis and other forms of degenerative diseases of the cartilage are altogether extremely prevalent. While genetic predisposition is uncontested, additional factors are involved too, including aging, obesity, and joint injuries. As for chondrodysplasias, the exact causes of these diseases and their pathogenetic mechanisms are still under study, and these research efforts are essential in order to fill major gaps that remain in the availability of efficient preventive, therapeutic, and curative treatments.

Notwithstanding clinical and scientific expertise, determination, rigor, and precision, successful research efforts require good experimental methods. Owing to its unique properties and those of its tissue matrix, the chondrocyte often poses technical challenges to the investigator and has therefore forced the design of specific methods and the tailoring of otherwise widely applied standard and cutting-edge approaches. In this book, investigators at the forefront of chondrocyte biology and pathology research have contributed a series of chapters explaining, step-by-step, a variety of methods that they have tailored for their successful research projects. Our hope is that the sharing of these methods will empower investigators worldwide by allowing them to readily and successfully incorporate new, cutting-edge approaches in their research project and thereby tackle and answer important questions. This book is divided into three sections, each comprising a handful of chapters.

The first section describes experimental models to study chondrocytes. Opening this section, Somoza and Welter explain the method that they have optimized to isolate chondrocytes from human cartilage and to culture the cells either in monolayer or in three-dimensional conditions that better preserve the differentiated phenotype of the cells (Chapter 1). Next, Uzielene, Mobasheri, and colleagues provide a method for primary human chondrocyte cultures that maintains the natural pericellular matrix of the cells, an advantage to keep the chondrocyte phenotype, and they explain how they isolate these chondron entities from adult osteoarthritic or normal cartilage (Chapter 2). As genetically modified mice offer the precious advantage of allowing chondrocytes to be studied in vivo and also of modeling human diseases, Kanakis, Bou-Gharios, and collaborators review the various *Cre* recombinase transgenes and knockin alleles that have been generated and validated by many teams to delete, overexpress, or otherwise alter genes exclusively in chondrocytes and, in some cases, only in specific subsets of the cells, such as articular chondrocytes (Chapter 3). Even when mouse models are available, it is often beneficial to extract chondrocytes and culture them in vitro to better address mechanistic questions. To this aim, Haseeb and Lefebvre outline their protocol to establish primary chondrocyte cultures from mouse growth plate and articular cartilage along with a discussion of the limitations of previously described methods (Chapter 4).

The second section is devoted to in vivo assays. Tata, Merceron, and Schipani share their expertise on histological and immunohistochemical techniques for fetal growth plate cartilage (Chapter 5). With more in situ methods, de Charleroy, Haseeb, and Lefebvre report on a well-validated method of mRNA in situ hybridization of adult mouse skeletal tissue sections (Chapter 6). The method was developed to preserve and readily detect mRNAs in cartilage as well as in bone and other adjacent tissues following sample fixation and quick but thorough demineralization. Endisha and Kapoor provide a complementary protocol to specifically detect microRNAs by in situ hybridization of paraffin-embedded human articular cartilage and mouse knee joints (Chapter 7). The next chapters are protocols to extract specific RNAs or other cartilage components from cartilage for quantitative and qualitative assays. Kikani and Lui teach us how to use laser capture microdissection to extract mouse growth plate cartilage samples for various downstream assays (Chapter 8). Zheng and McAlinden reveal their optimized technique to directly isolate quality RNA from articular cartilage tissue (Chapter 9), while Peffers, Cremers, and Welting communicate their method to profile small nucleolar RNA expression in cartilage (Chapter 10).

The third section describes methods to characterize the chondrocyte phenotype with high-throughput or discrete assays on cells maintained in vivo, just isolated from their tissue (ex vivo), or cultured in vitro. Chiba, Lotz, Asahara, and colleagues explain in depth high-throughput sequencing assays for microRNA expression profiling, target identification, and validation in chondrocytes (Chapter 11), and Yamakawa, Ohba, and colleagues describe in detail high-throughput chromatin immunoprecipitation assays to characterize epigenetic and transcriptional regulation of gene expression in chondrocytes (Chapter 12). Ramos and Mobasheri then report on their technique to assess the roles of microRNAs in the epigenetic regulation of autophagy in chondrocytes (Chapter 13). The book concludes with two chapters dedicated to the analysis of mitochondria, powerhouse organelles in chondrocytes as in other cells. Wang, Chen, and Bryan reveal their methods to assay mitochondrial biogenesis, activity, and DNA isolation in chondrocytes (Chapter 14), and Ansari and Haqqi inform us on their procedure to assess the chondrocyte status by immunofluorescence-mediated localization of Parkin relative to mitochondria (Chapter 15).

Finally, we would like to thank all the authors as well as Dr. John Walker, series editor, for all the efforts, enthusiasm, and professionalism that they demonstrated in sharing their expertise and thereby in helping us to assemble this first edition of *Chondrocytes: Methods and Protocols*. We hope that the book will be warmly acclaimed by experts as well as newcomers in the cartilage field and that it will spark new research ideas and efforts to move the field forward and solve many of the mysteries that are still pending regarding cartilage biology, disease mechanisms, and potential treatments.

Tariq M. Haqqi
Department of Anatomy and Neurobiology
Northeast Ohio Medical University
Rootstown, OH, USA

Véronique Lefebvre
Translational Research Program in Pediatric Orthopaedics
Department of Surgery/Division of Orthopaedic Surgery
Children's Hospital of Philadelphia
Abramson Research Center
Philadelphia, PA, USA

Contents

Contributors

MOHAMMAD ALHASHMI • *Department of Musculoskeletal and Ageing Science, Institute of Life Course and Medical Sciences, University of Liverpool, Liverpool, UK*

MOHAMMAD Y. ANSARI • *Department of Anatomy and Neurobiology, Northeast Ohio Medical University, Rootstown, OH, USA*

HIROSHI ASAHARA • *Department of Systems BioMedicine, Tokyo Medical and Dental University (TMDU), Tokyo, Japan; Department of Molecular Medicine, The Scripps Research Institute, La Jolla, CA, USA*

EIVA BERNOTIENE • *Department of Regenerative Medicine, State Research Institute Centre for Innovative Medicine, Vilnius, Lithuania*

GEORGE BOU-GHARIOS • *Department of Musculoskeletal and Ageing Science, Institute of Life Course and Medical Sciences, University of Liverpool, Liverpool, UK*

LIANG-YU CHEN • *VA San Diego Healthcare System, San Diego, CA, USA*

TOMOKI CHIBA • *Department of Systems BioMedicine, Tokyo Medical and Dental University (TMDU), Tokyo, Japan*

ANDY CREMERS • *Laboratory for Experimental Orthopedics, Department of Orthopedic Surgery, Maastricht University Medical Centre, Maastricht, The Netherlands*

CHARLES DE CHARLEROY • *Division of Orthopaedic Surgery, Department of Surgery, The Children's Hospital of Philadelphia, Philadelphia, PA, USA*

JAROSLAV DENKOVSKIJ • *Department of Regenerative Medicine, State Research Institute Centre for Innovative Medicine, Vilnius, Lithuania*

HELAL ENDISHA • *Arthritis Program, Krembil Research Institute, University Health Network, Toronto, ON, Canada; Department of Laboratory Medicine and Pathobiology, University of Toronto, Toronto, ON, Canada*

TARIQ M. HAQQI • *Department of Anatomy and Neurobiology, Northeast Ohio Medical University, Rootstown, OH, USA*

ABDUL HASEEB • *Division of Orthopaedic Surgery, Department of Surgery, The Children's Hospital of Philadelphia, Philadelphia, PA, USA*

HIRONORI HOJO • *Center for Disease Biology and Integrative Medicine, Graduate School of Medicine, The University of Tokyo, Tokyo, Japan; Department of Bioengineering, Graduate School of Engineering, The University of Tokyo, Tokyo, Japan*

YOSHIAKI ITO • *Department of Systems BioMedicine, Tokyo Medical and Dental University (TMDU), Tokyo, Japan*

URSULE KALVAITYTE • *Department of Regenerative Medicine, State Research Institute Centre for Innovative Medicine, Vilnius, Lithuania*

IOANNIS KANAKIS • *Department of Musculoskeletal and Ageing Science, Institute of Life Course and Medical Sciences, University of Liverpool, Liverpool, UK*

MOHIT KAPOOR • *Arthritis Program, Krembil Research Institute, University Health Network, Toronto, ON, Canada; Department of Laboratory Medicine and Pathobiology, University of Toronto, Toronto, ON, Canada; Department of Surgery, University of Toronto, Toronto, ON, Canada*

CRAIG KEENAN • *Department of Musculoskeletal and Ageing Science, Institute of Life Course and Medical Sciences, University of Liverpool, Liverpool, UK*

BIJAL KIKANI • *Section on Growth and Development, Eunice Kennedy Shriver National Institute of Child Health and Human Development, National Institutes of Health, Bethesda, MD, USA*

RYOTA KURIMOTO • *Department of Systems BioMedicine, Tokyo Medical and Dental University (TMDU), Tokyo, Japan*

VÉRONIQUE LEFEBVRE • *Division of Orthopaedic Surgery, Department of Surgery, The Children's Hospital of Philadelphia, Philadelphia, PA, USA*

KE LIU • *Department of Musculoskeletal and Ageing Science, Institute of Life Course and Medical Sciences, University of Liverpool, Liverpool, UK*

RU LIU-BRYAN • *VA San Diego Healthcare System, San Diego, CA, USA; Department of Medicine, University of California San Diego, La Jolla, CA, USA*

MARTIN LOTZ • *Department of Molecular Medicine, The Scripps Research Institute, La Jolla, CA, USA*

JULIAN C. LUI • *Section on Growth and Development, Eunice Kennedy Shriver National Institute of Child Health and Human Development, National Institutes of Health, Bethesda, MD, USA*

TAKAHIDE MATSUSHIMA • *Department of Systems BioMedicine, Tokyo Medical and Dental University (TMDU), Tokyo, Japan*

AUDREY MCALINDEN • *Department of Orthopaedic Surgery, Washington University School of Medicine, St Louis, MO, USA; Department of Cell Biology, Washington University School of Medicine, St. Louis, MO, USA; Shriners Hospital for Children—St Louis, St Louis, MO, USA*

CHRISTOPHE MERCERON • *Departments of Orthopedic Surgery, Medicine and Cell and Developmental Biology, University of Michigan School of Medicine, Ann Arbor, MI, USA*

ALI MOBASHERI • *Department of Regenerative Medicine, State Research Institute Centre for Innovative Medicine, Vilnius, Lithuania; Research Unit of Medical Imaging, Physics and Technology, Faculty of Medicine, University of Oulu, Oulu, Finland; Departments of Orthopedics, Rheumatology and Clinical Immunology, University Medical Center Utrecht, Utrecht, The Netherlands; Centre for Sport, Exercise and Osteoarthritis Research Versus Arthritis, Queen's Medical Centre, Nottingham, UK*

RYO NAKAMICHI • *Department of Molecular Medicine, The Scripps Research Institute, La Jolla, CA, USA*

SHINSUKE OHBA • *Center for Disease Biology and Integrative Medicine, Graduate School of Medicine, The University of Tokyo, Tokyo, Japan; Department of Cell Biology, Institute of Biomedical Sciences, Nagasaki University, Nagasaki, Japan*

MANDY J. PEFFERS • *Department of Musculoskeletal Biology, Institute of Life Course and Medical Sciences, University of Liverpool, Liverpool, UK*

BLANDINE POULET • *Department of Musculoskeletal and Ageing Science, Institute of Life Course and Medical Sciences, University of Liverpool, Liverpool, UK*

YOLANDE F. M. RAMOS • *Department of Biomedical Data Sciences, Section Molecular Epidemiology, Leiden University Medical Center, Leiden, The Netherlands*

LORENZO RAMOS MUCCI • *Department of Musculoskeletal and Ageing Science, Institute of Life Course and Medical Sciences, University of Liverpool, Liverpool, UK*

ERNESTINA SCHIPANI • *Departments of Orthopedic Surgery, Medicine and Cell and Developmental Biology, University of Michigan School of Medicine, Ann Arbor, MI, USA*

RODRIGO A. SOMOZA • *Department of Biology, Skeletal Research Center, Case Center for Multimodal Evaluation of Engineered Cartilage, Case Western Reserve University, Cleveland, OH, USA*

ZACHARY TATA • *Departments of Orthopedic Surgery, Medicine and Cell and Developmental Biology, University of Michigan School of Medicine, Ann Arbor, MI, USA*

ILONA UZIELIENE • *Department of Regenerative Medicine, State Research Institute Centre for Innovative Medicine, Vilnius, Lithuania*

RAMINTA VAICIULEVICIUTE • *Department of Regenerative Medicine, State Research Institute Centre for Innovative Medicine, Vilnius, Lithuania*

YUN WANG • *VA San Diego Healthcare System, San Diego, CA, USA; Department of Medicine, University of California San Diego, La Jolla, CA, USA*

JEAN F. WELTER • *Department of Biology, Skeletal Research Center, Case Center for Multimodal Evaluation of Engineered Cartilage, Case Western Reserve University, Cleveland, OH, USA*

TIM J. M. WELTING • *Laboratory for Experimental Orthopedics, Department of Orthopedic Surgery, Maastricht University Medical Centre, Maastricht, The Netherlands*

AKIRA YAMAKAWA • *Center for Disease Biology and Integrative Medicine, Graduate School of Medicine, The University of Tokyo, Tokyo, Japan; Department of Bioengineering, Graduate School of Engineering, The University of Tokyo, Tokyo, Japan*

HONGJUN ZHENG • *Department of Orthopaedic Surgery, Washington University School of Medicine, St Louis, MO, USA*

Chapter 1

Isolation of Chondrocytes from Human Cartilage and Cultures in Monolayer and 3D

Rodrigo A. Somoza and Jean F. Welter

Abstract

Chondrocytes are the only cell type in cartilage. The dense cartilage extracellular matrix surrounding the chondrocytes makes isolating these cells a complex and lengthy task that subjects the cells to harsh conditions. Protocols to isolate expand and maintain these cells have been improved over the years, providing ways to obtain viable cells for tissue engineering and clinical applications. Here we describe a method to obtain populations of chondrocytes that are able to expand and maintain a native-like phenotype.

Key words Human chondrocytes, Cartilage ECM, Sequential digestion

1 Introduction

Articular cartilage structure and function can easily be compromised due to injury, aging, or pathological conditions, resulting in significant reduction of quality of life. This is a major clinical challenge that often leads to osteoarthritis and, ultimately, total joint replacement surgery [1]. Due to its avascular nature, articular cartilage has little or no capacity to regenerate itself [2]. Today, only a few cell-based products have the potential to replace or repair injured cartilage, such as autologous chondrocyte implantation [3] and tissue engineering advances; therefore, the use of chondrocyte is of paramount importance in the design of regenerative-based approaches.

Only about 2–10% of the cartilage tissue volume consists of chondrocytes, which makes it one of the tissues with the lowest cell densities [4, 5]. Chondrocytes are highly specialized cells localized within the dense cartilage extracellular matrix (ECM) in matrix cavities known as lacunae [5]. In order to design and implement suitable approaches for cartilage repair/regeneration in a clinical setting there is a need to obtain large populations of viable chondrocytes. The isolation of chondrocytes from their dense ECM is

Tariq M. Haqqi and Véronique Lefebvre (eds.), *Chondrocytes: Methods and Protocols*, Methods in Molecular Biology, vol. 2245, https://doi.org/10.1007/978-1-0716-1119-7_1,

particularly important for upstream therapeutic and tissue-engineered applications. Success in doing so will depend on the successful implementation of techniques for harvest, isolation, expansion, and phenotype maintenance of in vitro cultured chondrocytes. Protocols to isolate expand and maintain these cells have been improved significantly over the past 50+ years [6].

For the purposes of this chapter, we will assume for the primary protocol that articular or costal cartilage is used as the source material. Other cartilage types (nasal, auricular) are treated similarly although the enzyme cocktail and protocol for isolation may vary (*see* **Note 4**). Generally, for isolated cartilage, cell viability decreases with time of the tissue ex vivo. For this reason, these samples will generally be from surgical tissue, so an IRB-approved protocol and a committed orthopedic surgeon and team will be a prerequisite for success. However, if isolated (closed joint capsule), handled, and transported properly, tissue from a tissue procurement service can also yield viable cells.

2 Materials

1. Appropriate personal protective equipment (PPE).
   (a) Lab coat.
   (b) Gloves.
   (c) Eye protection.
2. Biological safety cabinet.
3. Cell culture incubator.
4. Inverted microscope.
5. Tabletop centrifuge with rotor for 50 ml tubes.
6. Pipette-Aid.
7. Sterile 50 ml, polypropylene tubes.
8. Sterile plastic disposable serological pipettes 10 and 25 ml.
9. Wide-orifice pipette tips.
10. T175 flasks.
11. 100 mm petri culture dishes
12. 20 ml syringes with 18 gauge needles
13. Sterile cell strainer, 70 μm pore size.
14. Filter unit with receiver 0.2 μm.
15. Sterile forceps (2).
16. #10 sterile scalpel or #10 scalpel blades + blade holders
17. Ethyl alcohol: 95%, 190 proof.

18. Dulbecco's phosphate buffered saline (DPBS), without calcium and magnesium 1×.
19. Hanks' balanced salt solution (Sigma Chemical Co.) 1×.
20. DMEM-LG: Dulbecco's modified Eagle's medium with 1 g/l glucose,
21. DMEM-HG: Dulbecco's modified Eagle's medium with 4 g/l glucose.
22. Antibiotic–antimycotic, 100× stock solution 10,000 U of penicillin G, 10 mg of streptomycin, and 2.5 μg of amphotericin B per ml in water.
23. FBS: fetal bovine serum.
24. Sodium pyruvate: 100× stock solution 100 mM in water.
25. Nonessential amino acids: 100× (Invitrogen).
26. ITS+: recombinant human insulin 1.0 mg/ml, human transferrin 0.5 mg/ml, selenite 6.7×10^{-4} mg/ml, linoleic acid 0.5 mg/ml, and bovine serum albumin 100 mg/ml (R&D systems).
27. Dexamethasone stock solution: 10^{-4} M in ethanol.
28. Ascorbic acid-2 phosphate stock solution: 800 mM in water (Wako Chemicals).
29. Transforming growth factor β1: 20 μg/ml in sterile 4 mM HCl containing 1 mg/ml bovine serum albumin.
30. Complete chondrocyte medium: DMEM-HG, 1% antibiotic–antimycotic solution, 1% sodium pyruvate, 1% nonessential amino acids (Invitrogen), 1% ITS+, 10^{-7} M dexamethasone, 80 mM ascorbic acid-2 phosphate, transforming growth factor β-1 (10 ng/ml).
31. Serum supplemented DMEM: DMEM-LG with 10% FBS.
32. Trypsin: 2.0 mg/ml in PBS.
33. Testicular hyaluronidase: 2.0 mg/ml (330 Units/ml) in DMEM (Sigma Chemical Co.), prepare fresh.
34. Type II collagenase: 2.0 mg/ml (250 Units/ml) in DMEM (Worthington Biochemical Corporation), prepare fresh.
35. Trypsin–EDTA: 0.05% trypsin in 0.53 mM EDTA.
36. Milli-Q water (18.2 MΩ·cm).
37. Trypan Blue: 0.4% in PBS.
38. Polypropylene v-bottom 96-well plates and lids.

3 Methods

3.1 Chondrocyte Isolation

1. General Safety: All human tissues should be considered to be potentially contaminated with bloodborne pathogens and so should be handled using Standard Precautions, including proper hand hygiene, use of PPE (gloves, lab coat, face shield), and proper disposal of waste, including liquids and sharps used in the processing of the tissue [7].
2. The fresh harvested cartilage sample is immediately placed into a sterile 50 ml centrifuge tube filled with DMEM-LG, supplemented with 1% antibiotic/antimycotic solution by the Operating Room team. The cartilage is kept at 4 °C and then transported to the lab as expediently as possible, in these tubes, in a Styrofoam box. Isolation steps should be performed within 48 h.
3. Fresh tissue from a tissue procurement service should always be maintained in a saline solution and kept at 4 °C during transport and storage. In our experience we have been able to recover viable chondrocytes even from samples that have been stored for 60 days before harvesting. Although long storage of cartilage tissue is not ideal for chondrocyte isolation, it provides flexibility that sometimes is needed to process this samples.
4. When planning the isolation, be aware that digestion protocols include an overnight incubation.

3.2 Isolation of Chondrocytes with Sequential Digestion (Recommended for Articular Cartilage)

1. Prepare complete chondrocyte medium.
2. Prepare enzyme solutions for sequential treatment. All enzymatic digestion solutions should be sterile-filtered (0.2 μm) prior to use.
   (a) Trypsin.
   (b) Testicular hyaluronidase (freshly prepared).
   (c) Type II collagenase (freshly prepared).
   (d) Trypsin (0.05%) and EDTA.
3. If applicable, remove perichondrium or any repair tissue.
4. Cut 1 mm thick full-thickness slices of cartilage down to the bone using sharp dissection (*see* **Notes 1** and **2**).
5. Cut the cartilage slices into pieces approximately 1 mm^3.
6. Transfer the pieces into a sterile 100 ml glass flask with a stir-bar and flood with 20 ml enzymatic digestion solution (same volume for all enzyme solutions).
7. Place on a stirrer and digest at 37 °C in a humidified atmosphere of 5% CO_2 and 95% air using the following enzymatic sequence.

   (a) Trypsin (30 min).
   (b) Testicular hyaluronidase (60 min).
   (c) Collagenase type II (overnight) (*see* **Note 3**).
8. Alternatively, pieces can be placed on a 100 mm petri culture dish with 10 ml enzymatic digestion solution and placed on an orbital shaker.
9. Remove supernatant and add 10 ml FBS to it to stop, save this.
10. Digest the remaining undigested cartilage with 0.25% collagenase type II and 0.5% trypsin for 2–3 h at humidified atmosphere (*see* **Note 4**).
11. Remove the supernatant and stop the digestion with 10 ml FBS.
12. Either pool the supernatants or process separately, as preferred (*see* **Note 5**).
13. Strain through the sterile 70 μm pore size cell strainer to remove debris.
14. Centrifuge at 260 × g for 10 min.
15. Wash the pellet twice with DMEM-LG supplemented with 10% FBS (*see* **Note 6**).
16. Resuspend in complete chondrocyte medium.
17. Determine cell number and viability using a hemocytometer and Trypan Blue vital dye. Plate at 200,000 cells per 10 cm petri culture dishes in complete chondrocyte medium (*see* **Note 7**).
18. Incubate cells in cell-culture incubator at 37 °C and 5%CO_2.

3.3 Propagation and Routine Maintenance of Chondrocytes in 2D

1. Examine cultures under inverted microscope daily, note cell growth or any signs of contamination.
2. During the first week, individual cell colonies should be observed (Fig. 1).
3. Change the medium every third or fourth day by aspirating and discarding spent medium, and replace with the 25 ml of fresh complete prewarmed chondrocyte medium per flask.
4. Recap the flask(s), and return to the incubator.
5. Dispose of spent culture medium into 10% bleach.
6. Subculture is carried out when culture flasks are 80–90% confluent. Primary cultures usually take 10–16 days before they are ready to be subcultured. If cells are ready to passage, trypsinize cells according to the trypsinization protocol, below, then:
7. Resuspend the cell pellet(s) in chondrocyte medium. For each T175 flask, resuspend cells in 3 ml of complete chondrocyte medium.

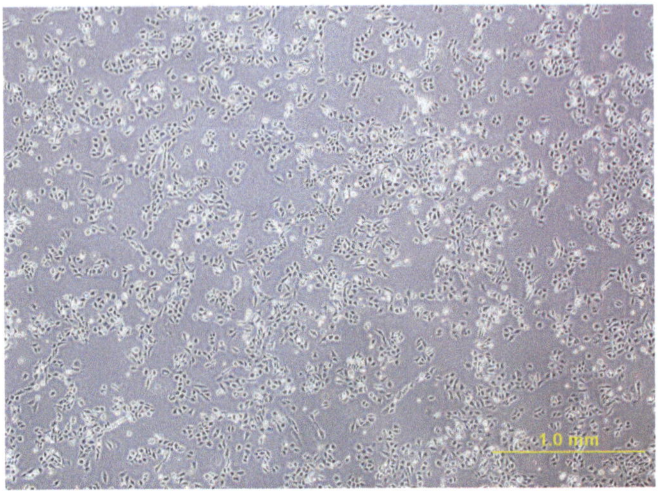

Fig. 1 Human chondrocytes isolated from the femoral condyle of a human knee cartilage. Chondrocytes were isolated based on the above protocol and expanded for 7 days. During this stage (in the presence of FBS), the cells divide rapidly and form colonies

8. Thoroughly mix the suspension by pipetting up and down.
9. Remove an aliquot (20 μl) and dilute 1:1 with 0.4% Trypan Blue. Count cells and calculate cell density and the total number of cells using a hemocytometer.
10. Adjust cell concentration to 1×10^6 cells/ml.
11. Add 1 ml of the cell suspension to each T175.
12. Add 25 ml complete chondrocyte medium.
13. Cap the flask(s) and return to the incubator.

3.4 Trypsinization of Monolayer-Cultured Chondrocytes

The purpose of this procedure is to remove the chondrocytes from the vessel surface for use, propagation, or cryopreservation. As with the other protocols in this chapter, Standard Precautions apply to any human-derived cultures. All applicable PPE and engineering controls should be used.

The attachment of cells to each other and to the culture substrate is mediated by cell surface glycoproteins and Ca^{2+} and Mg^{2+} ions. Proteins derived from serum also facilitate cell adhesion. To detach the cells from flask, remove culture medium, wash with DPBS, and expose to trypsin–EDTA.

1. Change the medium on the day before trypsinization.
2. Thaw a sufficient number of 50 ml sterile aliquot of trypsin–EDTA in a water bath at 37 °C. Discard unused portion at the end of the procedure.
3. Prepare growth chondrocyte medium or freezing medium as needed.

4. Warm DPBS or Hanks' balanced salt solution.
5. If there are a large number of flasks to trypsinize, do only five flasks at a time, as leaving the trypsin–EDTA on the cells for more than 10 minutes decreases viability.
6. For a T175 flask, add 20 ml of the prewarmed DPBS to the side of the flask opposite the cells (to avoid dislodging the cells).
7. Rinse the cell layer by gently swirling the DPBS on the cell layer.
8. Discard rinse.
9. Add 8 ml of the prewarmed trypsin–EDTA to the side of the flask opposite the cells. Turn flask over to cover the cell layer completely. Incubate cells at 37 °C in incubator for 5 min.
10. Check the progress of detachment by examining flasks under the inverted microscope to ensure that the cells have completely detached prior to recovery of the cells. Do not force the cells to detach before they are ready to do so, or clumping may result (*see* **Note 8**).
11. Add 8 ml of complete chondrocyte medium to each flask to stop the action of the trypsin. Disperse cells by repeated pipetting or gentle agitation.
12. Transfer the suspension(s) to sterile 50 ml conical centrifuge tube(s). Rinse flask with 10 ml of DPBS and add this to the suspension.
13. Centrifuge at 180 × *g* for 5 min at 37 °C.
14. Aspirate the supernatant and resuspend the cell pellet in 5 ml of chondrocyte medium.
15. If multiple tubes are used, pool cell suspension in one tube. Count cells and then use; replate at 1×10^6 cells per T175 as needed.

3.5 Chondrocyte Culture in 3D Aggregate Culture

1. For use in aggregate culture, resuspend the trypsinized cells to a density 0.25×10^6 cells/ml in complete chondrocyte medium.
2. Pipet a 200 μl aliquot into each well of a sterile polypropylene v-bottom 96 well plate for 0.25×10^6 cells per well [8] (*see* **Notes 9** and **10**).
3. Centrifuge the plates at 500 × *g* for 5 min and then gently transfer to the incubator.
4. Aggregates will form within the first 24 h.
5. After 24 h, gently pipet the medium in the wells up and down a few times to detach the aggregates from the bottom of the wells, taking care not to disrupt the aggregates.

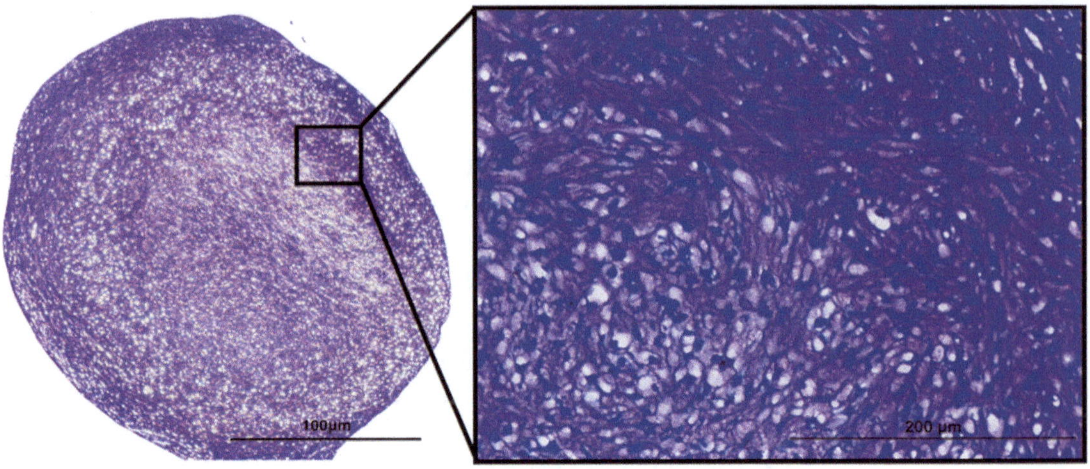

Fig. 2 Chondrocyte culture in 3D structures (aggregates). After 3 weeks in culture with chondrocyte medium, chondrocytes will produce cartilage extracellular matrix (Toluidine Blue stain) and will be positioned within lacunae inside the matrix

6. Change the medium per experimental protocol, but no less than every 3 days. Care should be taken not to aspirate the aggregates when changing the medium (*see* **Note 11**).
7. Aggregates can be harvested after 3 weeks. At this stage, chondrocytes inside lacunae will be observed within the 3D cartilage ECM (Fig. 2).
8. To harvest the aggregates, a wide-orifice pipette tip will usually allow for the aggregates to be picked up and transferred for upstream analyses.

4 Notes

1. Articular cartilage (especially in the knee) has different weight-bearing zones, which have adapted to different physiological loading conditions, and that result in different tissue and chondrocyte properties. Therefore, isolating chondrocytes specifically from different weight-bearing areas may be of research interest. The same isolation protocol can be used for either zone (weigh-bearing or non–weight-bearing) (Fig. 3).
2. When cutting slices of cartilage, it is important not to harvest tissue from the bone-cartilage interface. The presence of vascular channels in subchondral bone may contaminate chondrocyte cultures with other cell types.
3. Depending on the age of the cartilage donor, the incubation time with collagenase may have to be adjusted. Younger cartilage (neonatal to infant) may only need 4–6 h to be completely digested, in which case the reaction needs to be stopped. In our

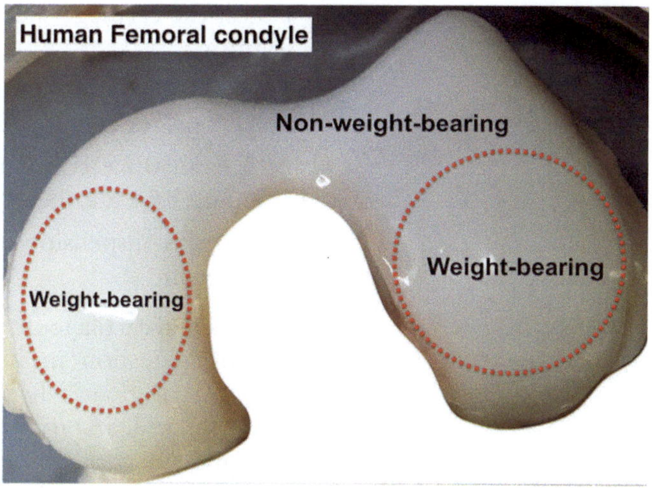

Fig. 3 Differential weight-bearing areas in the femoral condyle cartilage. Chondrocytes can be differentially isolated from weight-bearing and non–weight-bearing zones in articular cartilage [9]

experience, neonatal cartilage is completely digested after 4 h. In contrast, adult cartilage needs an overnight digestion.

4. Depending on the subtype of human cartilage (auricular, costal, nasoseptal, and articular), different enzymatic digestion sequences may be used:
   (a) An alternate digestion protocol (from Bradham et al. [10]) for articular cartilage (all in DMEM-LG at 37 °C in a humidified atmosphere of 5% CO_2 and 95% air).
   - Digest in 2 mg/ml hyaluronidase (45 min).
   - Digest in 2 mg/ml trypsin (45 min).
   - Digest in 4 mg/ml collagenase II for 2–3 h.

   (b) A further alternate digestion protocol (from Islam et al. [11]) (all in F12 medium at 37 °C in a humidified atmosphere of 5% CO_2 and 95% air).
   - Digest in 0.15% (w /v) Pronase for 1 h.
   - Digest in 0.15% (w/v) *Clostridium histolyticum* collagenase overnight.

   (c) Another alternate digestion protocol.
   - While collecting the cartilage, digest in hyaluronidase 0.05% in PBS.
   - Transfer minced cartilage to 0.2% trypsin for 45 min at 37 °C for 45 min with stirring.
   - Discard the trypsin solution.

- Digest the cartilage slices in 10 ml of 0.2% collagenase solution.
- Retain this suspension in a sterile 50-ml polypropylene tube.
- Add fresh 0.2% collagenase solution to the cartilage and digest for another 45 min at 37 °C with stirring.
- Retain this suspension as well, and pool with the first suspension.

(d) For nasoseptal cartilage (from Aigner et al. [12]): Perform enzymatic digestion in 2 mg/ml collagenase type II, 0.1 mg/ml testicular hyaluronidase, and 0.15 mg/ml DNase (Fluka, USA) in DMEM-LG at 37 °C in a humidified atmosphere of 5% CO2 and 95% air for 18 h.

(e) For auricular cartilage: Perform enzymatic sequential digestion with the following.

- 0.1% trypsin (30 min).
- 0.1% hyaluronidase (60 min).
- 0.1% collagenase type II (overnight).
- 0.15 mg/ml DNase (optional).

(all in DMEM-LG at 37 °C in a humidified atmosphere of 5% CO_2 and 95% air)

5. To maximize cell yield, the remaining undigested material after the second digestion (**step 10** from isolation protocol, Subheading 3.2) can be plated on tissue culture plastic (plates/flasks) to allow outgrowth of any remaining cells.
6. Serum lots vary in their ability to support chondrocyte grown. It is therefore important to screen several lots and purchase a batch of the best-performing serum.
7. Although chondrocytes can be expanded well in serum-supplemented DMEM, after the first passage a change into a fibroblastic-like phenotype can be observed. To prevent this, a chondrocyte specific medium should be used. In our experience, chondrocyte medium (*see* Subheading 2) provides a good proliferative support and chondrocyte phenotype maintenance based on the observation that articular cartilage markers expression is maintained compared to chondrocytes expanded in serum-supplemented DMEM (Fig. 4).
8. Chondrocytes can produce and deposit large quantities of ECM, especially when they are at high density. In this case, trypsin will not work very efficiently. To improve detachment, Accutase® solution (Sigma-Aldrich) can be used. This solution contains proteolytic and collagenolytic enzymes that work well in detaching cells with high production of ECM. Accutase®

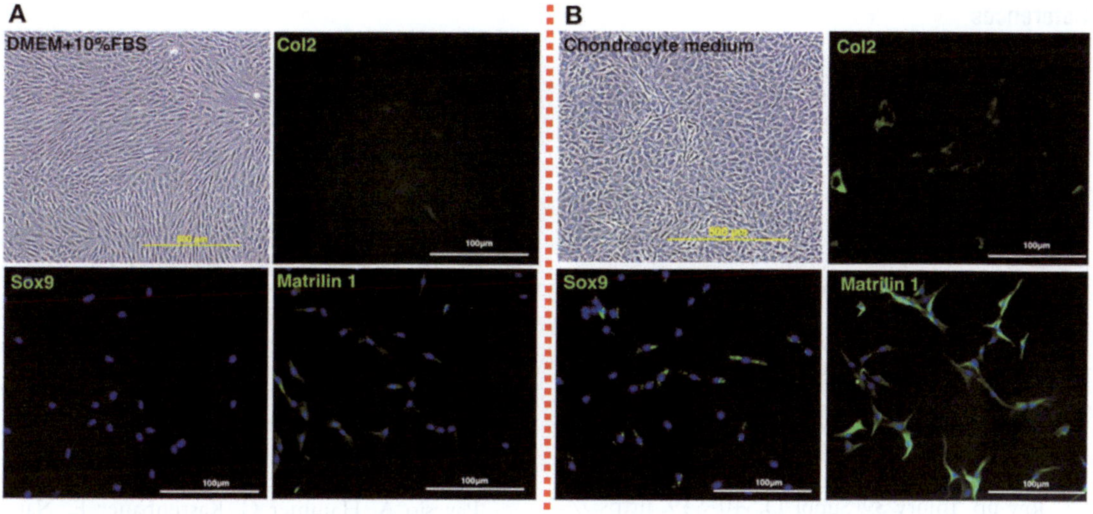

Fig. 4 Chondrocyte phenotype maintenance during culture. (**a**) Isolated chondrocytes can be efficiently expanded in the presence of FBS; however, these conditions may produce a loss of the molecular phenotype of native chondrocytes. This is evident by the acquisition of a fibroblastic-like morphology and low expression of chondrogenic markers. (**b**) After expansion in primary culture, the medium should be switched to complete chondrocyte medium (*see* Subheading 2). Under these conditions, chondrocytes continue to proliferate and maintain a native-like phenotype, such as a polygonal morphology and high expression of relevant chondrogenic markers [13]

may also be useful if trypsin-sensitive cell surface epitopes must be preserved (e.g., for immunophenotyping) [14].

9. Polypropylene 96-well plates can be difficult to find presterilized. They do survive autoclaving.
10. For aggregate culture in 96-well plates, it is useful to note that the medium in the peripheral wells evaporates faster than in the rest of the plates. Depending on the experiment, it may be useful to not seed these wells, but rather fill them with plain medium or DI water.
11. Alternate base media for chondrocyte culture include Ham's F12 or DMEM–F12 1:1 [15, 16].

Acknowledgments

This work was funded, in part, by grants from the National Institutes of Health P41 EB021911, R01 AR050208, and P01 AR053622.

References

1. McCormick F, Harris JD, Abrams GD, Frank R, Gupta A, Hussey K, Wilson H, Bach B Jr, Cole B (2014) Trends in the surgical treatment of articular cartilage lesions in the United States: an analysis of a large private-payer database over a period of 8 years. Arthroscopy 30(2):222–226. https://doi.org/10.1016/j.arthro.2013.11.001
2. Buckwalter J, Mankin H (1997) Instructional course lectures, The American Academy of Orthopaedic Surgeons—articular cartilage. Part II: degeneration and osteoarthrosis, repair, regeneration, and transplantation. J Bone Jt Surg 79(4):612–632
3. Brittberg M (2008) Autologous chondrocyte implantation—technique and long-term follow-up. Injury 39(Suppl 1):S40–S49. https://doi.org/10.1016/j.injury.2008.01.040
4. Alford JW, Cole BJ (2005) Cartilage restoration, part 1: basic science, historical perspective, patient evaluation, and treatment options. Am J Sports Med 33(2):295–306. https://doi.org/10.1177/0363546504273510
5. Hunziker EB, Quinn TM, Häuselmann HJ (2002) Quantitative structural organization of normal adult human articular cartilage. Osteoarthr Cart 10(7):564–572
6. Prockop DJ, Pettengill O, Holtzer H (1964) Incorporation of sulfate and the synthesis of collagen by cultures of embryonic chondrocytes. Biochim Biophys Acta 83:189–196. https://doi.org/10.1016/0926-6526(64)90034-5
7. OSHA (2019) Bloodborne pathogens. https://www.osha.gov/pls/oshaweb/owadisp.show_document?p_id=10051&p_table=STANDARDS
8. Penick KJ, Solchaga LA, Welter JF (2005) High-throughput aggregate culture system to assess the chondrogenic potential of mesenchymal stem cells. BioTechniques 39(5):687–691
9. Koo S, Gold GE, Andriacchi TP (2005) Considerations in measuring cartilage thickness using MRI: factors influencing reproducibility and accuracy. Osteoarthr Cart 13(9):782–789. https://doi.org/10.1016/j.joca.2005.04.013
10. Bradham DM, Horton WE Jr (1998) In vivo cartilage formation from growth factor modulated articular chondrocytes. Clin Orthop Relat Res 352:239–249
11. Islam N, Haqqi TM, Jepsen KJ, Kraay M, Welter JF, Goldberg VM, Malemud CJ (2002) Hydrostatic pressure induces apoptosis in human chondrocytes from osteoarthritic cartilage through up-regulation of tumor necrosis factor-alpha, inducible nitric oxide synthase, p53, c-myc, and bax-alpha, and suppression of bcl-2. J Cell Biochem 87(3):266–278. https://doi.org/10.1002/jcb.10317
12. Aigner J, Tegeler J, Hutzler P, Campoccia D, Pavesio A, Hammer C, Kastenbauer E, Naumann A (1998) Cartilage tissue engineering with novel nonwoven structured biomaterial based on hyaluronic acid benzyl ester. J Biomed Mater Res 42(2):172–181
13. Somoza RA, Correa D, Labat I, Sternberg H, Forrest ME, Khalil AM, West MD, Tesar P, Caplan AI (2018) Transcriptome-wide analyses of human neonatal articular cartilage and human mesenchymal stem cell-derived cartilage provide a new molecular target for evaluating engineered cartilage. Tissue Eng Part A 24(3–4):335–350. https://doi.org/10.1089/ten.TEA.2016.0559
14. Spaas JH, De Schauwer C, Cornillie P, Meyer E, Van Soom A, Van de Walle GR (2013) Culture and characterisation of equine peripheral blood mesenchymal stromal cells. Vet J 195(1):107–113. https://doi.org/10.1016/j.tvjl.2012.05.006
15. Webber RJ, Malemud CJ, Sokoloff L (1977) Species differences in cell culture of mammalian articular chondrocytes. Calcif Tissue Res 23(1):61–66
16. Jennings SD, Ham RG (1983) Clonal growth of primary cultures of human hyaline chondrocytes in a defined medium. Cell Biol Int Rep 7(2):149–159. https://doi.org/10.1016/0309-1651(83)90028-0

Chapter 2

Protocol for the Isolation of Intact Chondrons from Healthy and Osteoarthritic Human Articular Cartilage

Ilona Uzieliene, Jaroslav Denkovskij, Eiva Bernotiene, Ursule Kalvaityte, Raminta Vaiciuleviciute, Yolande F. M. Ramos, and Ali Mobasheri

Abstract

Chondrons are the main functional microanatomical units in cartilage, consisting of chondrocytes and the directly surrounding pericellular matrix (PCM). They have attracted attention as a more physiological and biomimetic in vitro model for evaluating chondrocyte function and metabolism as compared to single chondrocytes. Chondrons may be more suitable for in vitro studies than primary chondrocytes that have been isolated without PCM since their in situ and in vivo states remain intact: chondrocytes within their PCM do not undergo the rapid dedifferentiation that proliferating single chondrocytes undergo in culture. Therefore, chondrons may be a better model for studying chondrocyte biology and responses to pro-inflammatory and anti-inflammatory cytokines, growth factors and novel therapeutics. In this chapter, we present a concise and unified protocol for enzymatic isolation of intact chondrons from human articular cartilage and determination of their viability.

Key words Articular cartilage, Osteoarthritis, Chondrons, Chondrocytes, Extracellular matrix (ECM), Pericellular matrix (PCM)

1 Introduction

Articular cartilage is a highly specialized, avascular, aneural, and alymphatic load-bearing connective tissue with unique viscoelastic properties. Its principal function is to provide a smooth, lubricated surface for almost friction-free articulation in synovial joints and to facilitate the transmission of mechanical loads to the underlying subchondral bone. However, despite its mechanical resilience, articular cartilage is susceptible to trauma and degeneration in joint diseases such as osteoarthritis (OA) [1]. Cartilage is produced and maintained by highly specialized cells called chondrocytes. Primary chondrocytes, isolated from articular cartilage, are used extensively to study the cell biology and physiology of articular cartilage, understand the molecular and signaling mechanisms involved in the pathogenesis of OA and evaluate the mode of action

Tariq M. Haqqi and Véronique Lefebvre (eds.), *Chondrocytes: Methods and Protocols*, Methods in Molecular Biology, vol. 2245, https://doi.org/10.1007/978-1-0716-1119-7_2, © Springer Science+Business Media, LLC, part of Springer Nature 2021

of new drugs. However, despite the generic application of primary isolated chondrocytes in a preclinical research setting, these cells rapidly dedifferentiate and lose their differentiated characteristics upon expansion in absence of their natural pericellular matrix (PCM) [2].

1.1 Cartilage Structure

The integrity of cartilage structure is crucial for its biomechanical performance. The thickness of human articular cartilage varies from 1 to 7 mm, depending on joint site [3]. Larger weight-bearing joints (e.g., hip and knee joints) are covered by thicker cartilage than smaller, non–weight-bearing, joints. Cartilage is composed of a dense extracellular matrix (ECM) with a sparse distribution of chondrocytes. The ECM is primarily composed of water, collagens, proteoglycans, glycoproteins, and noncollagenous proteins [4]. Collagen type II (90–95%) is a major fibrillar collagen of hyaline cartilage, which forms dense fibrils and fibers intertwined with proteoglycan aggregates [5].

1.2 Chondrocytes

Chondrocytes are terminally differentiated cells. They derive from mesenchymal stromal cells (MSCs) and only make up around 1–5% of total cartilage volume [6]. The primary functions of chondrocytes include the production and physiological turnover of ECM, which consists of a meshwork of type II collagen, aggregating proteoglycans, and a number of other minor collagens (such as type IX and I) and glycoproteins [5]. As cartilage is an avascular tissue, chondrocytes function in a low oxygen environment with low metabolic turnover, which is a consequence of their low mitochondrial numbers [7]. For their nutrition chondrocytes rely only on diffusion from the articular surface. Isolation of human chondrocytes was first described by Manning et al. in 1967 and later modified and refined by other groups [8–10]. In articular cartilage, chondrocytes are surrounded by PCM which forms their natural niches that together with the enclosed cells are called chondrons [11].

1.3 Chondrons as a Niche for Chondrocytes

The original concept of the "chondron" was first introduced by Benninghoff back in 1925. He described it as an "inter-territorial" structure, consisting of single or multiple chondrocytes, surrounded by a rim and responsible for maintaining the homeostasis of the ECM [12]. Later, a more systematic perspective on chondron structure and function was introduced. The importance of further analysis of chondrons was perceived, but not until 1985 chondrons were isolated from hyaline cartilage by using serial low speed homogenization [13, 14]. The chondron is defined as the main cartilage functional and metabolic microanatomical unit, consisting of chondrocytes and the surrounding PCM. Figure 1a shows a schematic drawing of a chondron. The PCM of the chondron structure acts as a mediator between the ECM and the

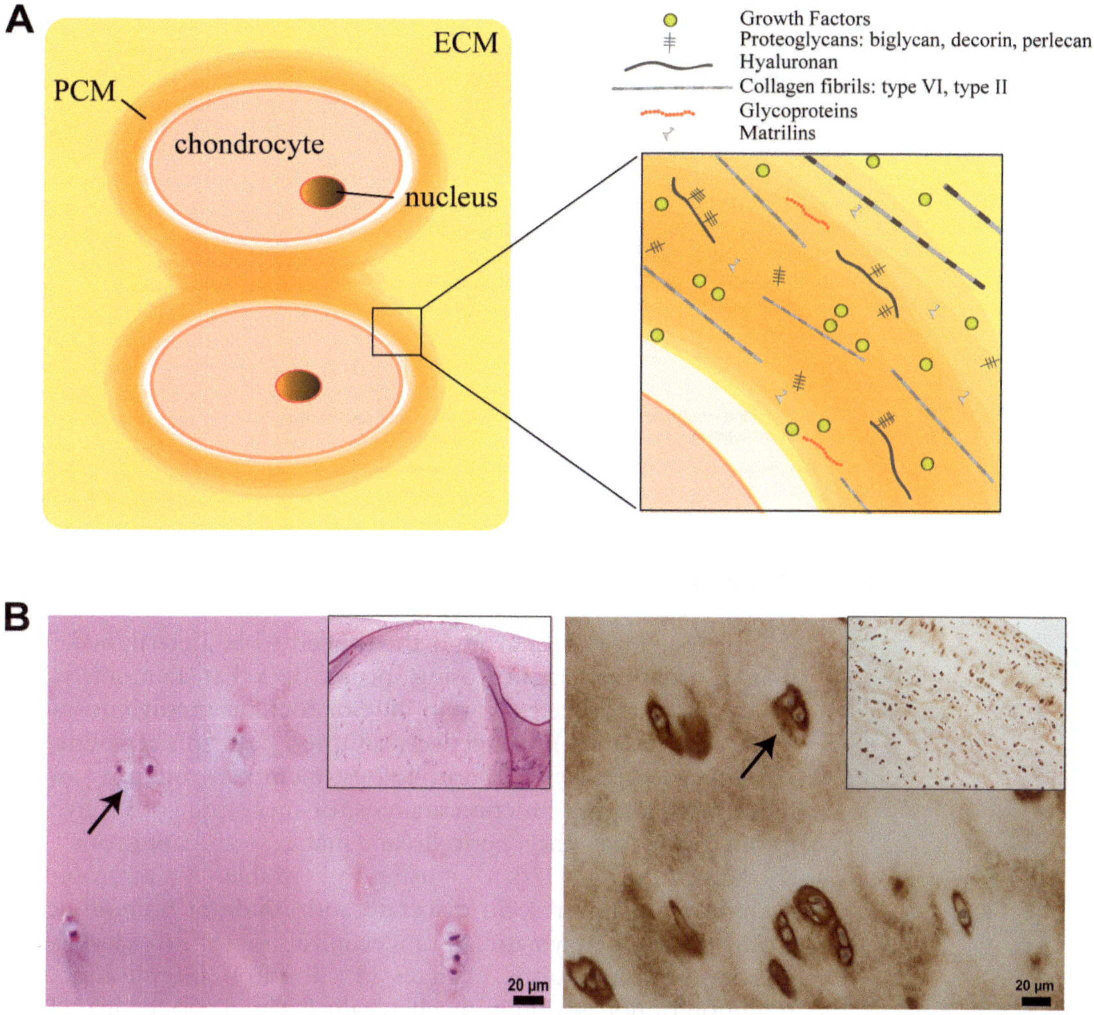

Fig. 1 The chondron. (**a**) Schematic overview of chondron, with indicated the pericellular matrix containing high levels of collagen fibrils (particularly type VI and type II collagen), hyaluronan, glycoproteins, matrilins, and proteoglycans (particularly biglycan, decorin, and perlecan) and growth factors (legend: ECM: extracellular matrix; PCM: pericellular matrix). (**b**) Paraffin-embedded articular cartilage from RAAK study [19] stained with eosin (left panel) and type VI collagen (right panel; ab6588, Abcam; 1:100 dilution), particularly localized in chondrons (counter staining with hematoxylin to visualize nuclei; arrow points at chondron)

chondrocytes, and contains high levels of type VI collagen, type II collagen, and proteoglycans, particularly biglycan, decorin, and perlecan [11, 14]. Type VI collagen forms fibril networks that are found only in the PCM, and is therefore often used as marker for chondrons (Fig. 1b; [15]). Due to its specific composition, the PCM is a key repository accountable for sequestration, storage, transportation, and degradation of various growth factors (e.g., fibroblast growth factor 2) [11].

For downstream analyses, after isolation of chondrons the number of viable cells and the number of chondrons that retain

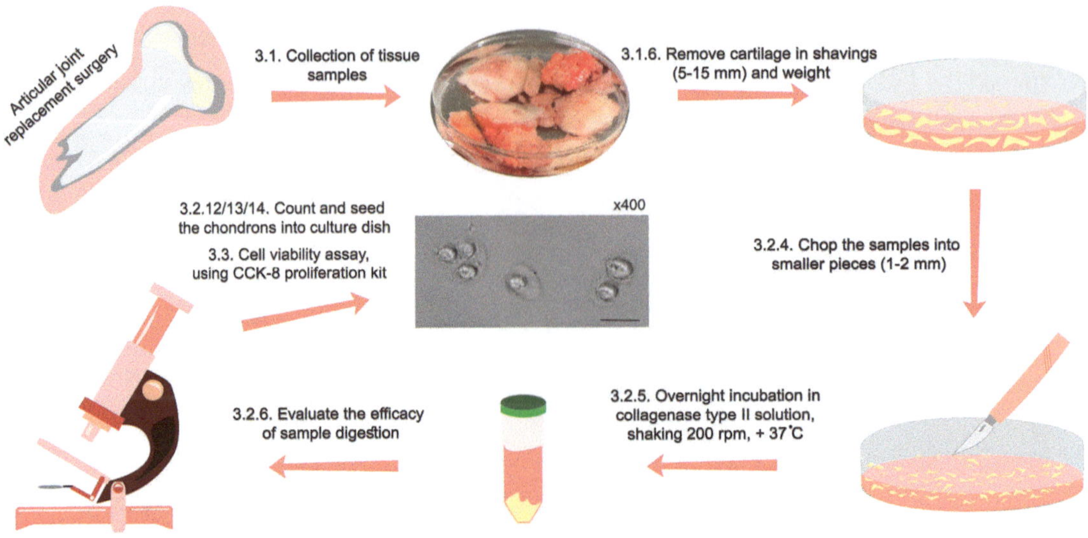

Fig. 2 Flowchart of chondron isolation protocol

the microenvironment around them need to be determined. Viability is often evaluated by using specific dyes that stain only viable or nonviable cells, (e.g., trypan blue or 5-chloromethylfluorescein diacetate (CMFDA), respectively). Other ways to determine cell viability are based on cell morphology, membrane integrity assessment methods, cell function analysis, or measurement of glycosaminoglycans in the pericellular matrix of chondrons and chondrocytes [16], using fixable fluoroprobes 5-chloromethylfluorescein diacetate and ethidium homodimer-1. Comparison of enzymatic and mechanical homogenization methods has revealed subtle changes in chondron morphology and retention of fibronectin in mechanically isolated chondrons. However, the average yield of enzyme-isolated chondrons was much greater than that reported for mechanically isolated chondrons [17].

Here we describe a step-by-step protocol for the enzymatic isolation of chondrons from osteoarthritic human articular cartilage (Fig. 2). The same method can be used for isolating chondrons from healthy cartilage. Furthermore, we describe the culture techniques used for the in vitro cultivation of chondrons, including determination of chondrocyte viability.

2 Materials

2.1 Reagents and Kits

1. Phosphate buffered saline (PBS) 0.9%, pH 7.4.
2. Clostridial Collagenase type II, lyophilized, 330 U/mg (Merck, cat. No C2-22).

Trypan blue solution.

Cell counting Kit-8 (CCK-8).

Fetal bovine serum (FBS) 10%.

2.2 Incubation Medium

Dulbecco's Modified Eagle Medium (DMEM) 1 g/L D-glucose, stable sodium pyruvate, and L-glutamine.

Penicillin–streptomycin (PS, 10,000 U/mL; 10,000 μg/mL; 1%).

Gentamicin (50 mg/mL; 1%).

2.3 Chondrogenic Medium

Dulbecco's Modified Eagle Medium (DMEM) 4.5 g/L D-glucose, stable sodium pyruvate, and L-glutamine.

Penicillin–streptomycin (PS) (10,000 U/mL; 10,000 μg/mL; 1%).

L-Proline (0.35 mM).

Ascorbic acid-2-phosphate (0.17 mM).

Insulin–transferrin–selenium (ITS).

Dexamethasone 10^{-7} M.

Fetal bovine serum (FBS) 10%.

3 Methods

Chondrons can be isolated from human articular cartilage tissue obtained from the knee joints of patients undergoing joint replacement surgery after receiving informed consent. Samples are collected into sterile dish and should be assembled for chondron isolation within 1–2 h.

3.1 Preparation of Human Articular Cartilage Samples

All procedures must be performed in the sterile environment of a laminar hood.

The samples during and after digestion are kept in a humidified 37 °C incubator with 5% CO_2 (later referred to as "incubator").

1. Prepare incubation medium, consisting of DMEM 1 g/L D-glucose and supplemented with 1% PS and 1% gentamicin. Prewarm it at 37 °C.
2. Transfer human articular joint tissues into a sterile culture glass dish Ø 10–15 cm (Figs. 2 and 3a), depending on the tissue sample size.
3. Wash the articular cartilage and subchondral bone twice with PBS, supplemented with 2% PS (room temperature).
4. Explore the articular tissues macroscopically, and select for the unaffected areas (i.e., with smooth surface) of cartilage tissue and remove residual soft tissues (*see* **Note 1**).

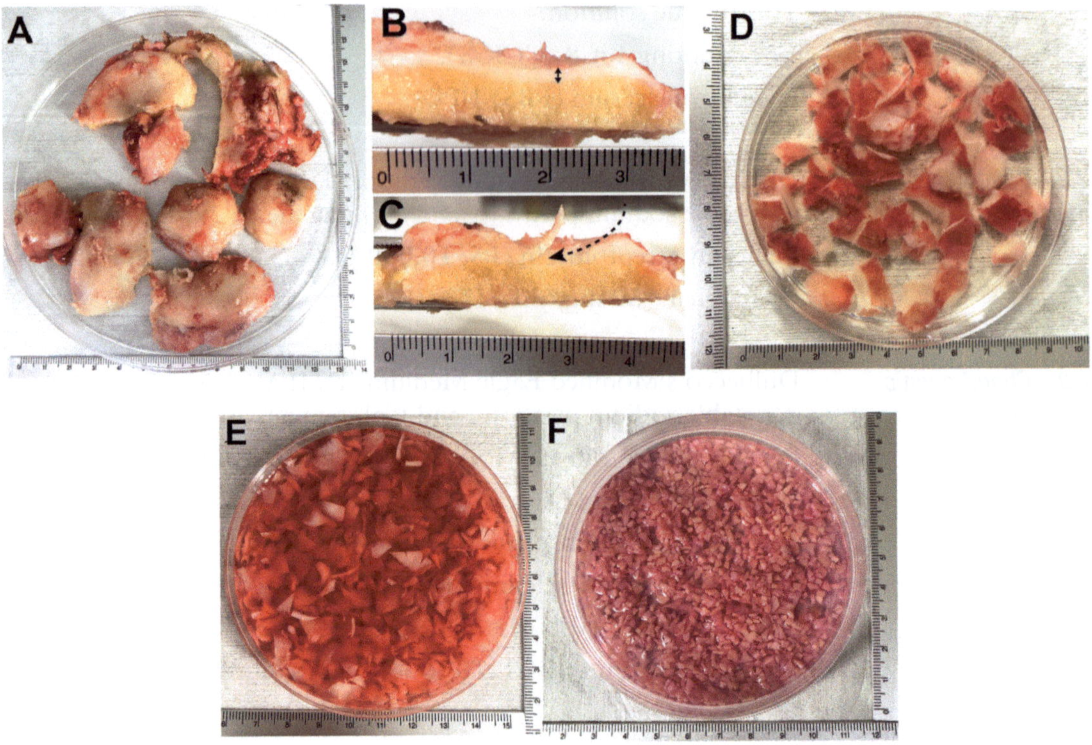

Fig. 3 Preparation of human articular cartilage samples for chondron isolation. (**a**) Human articular knee joint received from the operating theatre. (**b**) Human articular cartilage thickness (showed by arrow). (**c**) Removal of articular cartilage from subchondral bone. (**d**) Articular cartilage shavings after excision from the articular joint surface. (**e**) Articular cartilage shavings covered with medium, consisting of DMEM 1 g/L D-glucose supplemented with 2% PS and 1% gentamicin. (**f**) Articular cartilage samples chopped into small pieces after overnight incubation in DMEM medium

5. Carefully cut off the cartilage tissue layer in shavings, using a sterile scalpel, until the surface of the bone is reached (Fig. 3b, c and **Note 2**).
6. Cut the shavings into 5–15 mm size pieces and transfer them on scalpel into a new, preweighed Ø 10 cm culture dish (Fig. 3d).
7. Determine weight of collected samples.
8. Add incubation medium and ensure it covers all pieces throughout the remaining steps to avoid the tissue to dry out (Fig. 3e and **Note 3**).

3.2 Digesting Cartilage Samples for Chondron Isolation

1. Prepare collagenase type II solution in 50 mL tube, amount depending on the cartilage sample weight (240 U/mL of collagenase type II solution in 10 mL of DMEM 1 g/L D-glucose +2% PS per 1 g of sample).

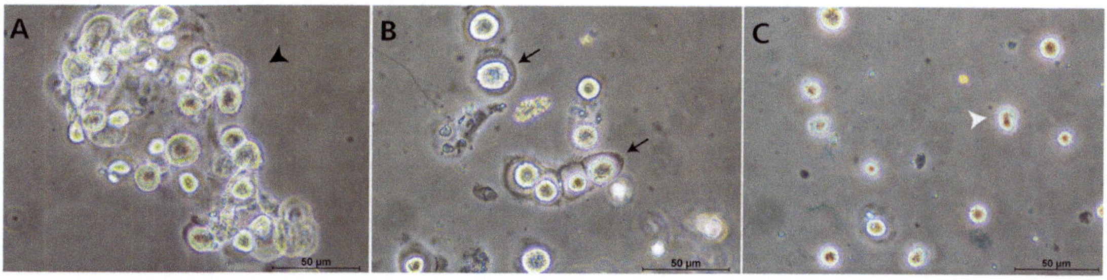

Fig. 4 Chondrons after cartilage digestion with collagenase type II. (**a**) Following overnight (approx. 16 h) incubation with collagenase II chondrons may not be fully released from the ECM and still reside in aggregates (black arrowhead). Additional incubation with collagenase is needed (magnification ×400). (**b**) Chondrons (indicated with an arrow) after isolation from cartilage using enzymatic digestion (magnification ×400). (**c**) Excessive enzymatic digestion, resulting in single chondrocytes (white arrowhead) released from chondrons

2. Filter-sterilize the collagenase type II solution using a 0.2 μm filter and prewarm it at 37 °C.
3. Remove the incubation medium using a 10 mL pipette.
4. Chop the cartilage shavings into smaller pieces (1–2 mm size; Fig. 3f).
5. Transfer chopped pieces into 50 mL tube with collagenase type II solution and leave them overnight (± 16 h) in a shaker (200 rpm speed, 37 °C incubator).
6. The following morning, collect 10 μL of tissue suspension and microscopically (200× magnification) evaluate the degree of sample digestion. In case chondrons are not clearly visible yet, leave the sample for an additional hour of incubation (*see* **Note 4**), periodically monitoring under microscope until desired purity/separation (*see* Fig. 4).
7. Place a cell strainer (70 μm pore size) on a new 50 mL tube, and carefully collect the collagenase-treated tissues (supernatant) using a 10 mL pipette. Pipette the cell suspension through the cell strainer to remove undigested cartilage fragments and debris.
8. Prepare chondrogenic medium and prewarm it at 37 °C.
9. Spin down chondrons by centrifugation (10 min at 500 × *g*).
10. Remove the supernatant after centrifugation and wash the pellet two times by adding 20 mL of incubation medium with 10% FBS (10 min at 500 × *g*).
11. Remove the supernatant after washing and add 4 mL of chondrogenic medium.
12. Count the viable chondrocytes in chondrons, using 7 μL Trypan Blue + 7 μL chondron suspension in hemocytometer camera and evaluate remaining debris/dead cell population.

13. Prepare a new Ø 10 cm culture dish and seed the chondrons according to viable chondrocyte number as counted before (approx. 150 k chondrocytes/dish; **Note 5**).
14. The next day, replate the chondron suspension in a new flask or plate for the separation of chondrons from chondrocytes: chondrocytes remain attached to the tissue culture plastic (*see* **Note 6**).

3.3 Analysis of Chondrocyte Viability in Chondrons

1. Prepare 4 mL of CCK-8 solution, according to the manufacturer's recommendation.
2. Mix 3 mL of CCK-8 solution with freshly isolated chondron suspension (50,000 chondrocytes/mL).
3. Place chondrons with CCK-8 solution into three wells of 12-well plate 1 mL/well.
4. Incubate the plate with chondrons for 3 h in an incubator.
5. Collect the medium from three different wells into 96-well plate (100 μL/well) and add remaining 1 mL of CCK-8 solution (prepared at the beginning) as control into separate well.
6. Measure the absorption at 450 nm using plate reader (spectrophotometer).
7. If needed, repeat the procedures after 1, 3, or more days to evaluate chondrocyte viability (*see* **Note 7**).

4 Notes

1. Directions to macroscopically select for unaffected cartilage regions have been described previously [18]. In short, inspect joint tissue for the following:
   (a) Cartilage color (unaffected cartilage is "off-white").
   (b) Surface integrity (determined by fibrillation and/or crack formation).
   (c) Cartilage thickness and hardness (upon sampling with a scalpel.
2. During the procedure of collecting cartilage shavings by cutting off the cartilage from joint tissues in Subheading 3.1, it is crucial to avoid collecting subchondral bone together with cartilage. Carefully select the areas and cut off exclusively the surface of cartilage.
3. Optional, following **step 8** and before continuing the protocol, cartilage shaves can be left overnight (±16 h) in an incubator. This may ensure more uniform collagenase digestion.
4. While monitoring chondron digestion degree after the overnight incubation, it is important not to leave the tissues too

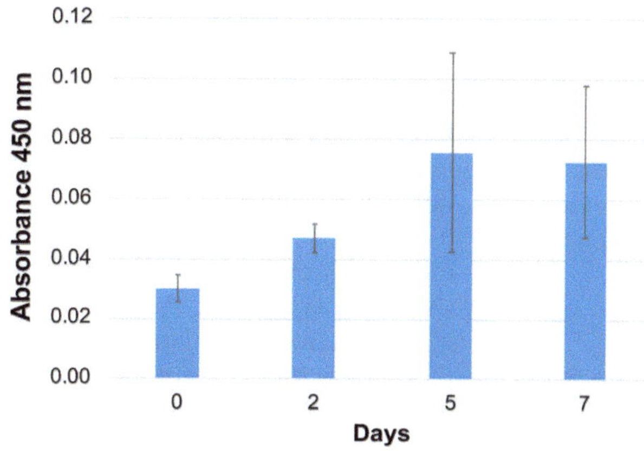

Fig. 5 Chondron viability test. Viability of chondrocytes in chondrons directly after isolation ($t = 0$) and following 2, 5, and 7 days in culture (absorbance of reduced formazan dye (CCK-8) generated by viable cells was analyzed at 450 nm)

long in collagenase type II solution (more than 20 h). This will further digest the pericellular matrix resulting in single chondrocytes. Therefore, if the cartilage pieces are not sufficiently digested yet after the overnight incubation and a lot of cartilage pieces are still visible, repeat digestion procedure, while visually monitoring the degree of sample digestion. After the supernatant with chondrons has been transferred to a new collection tubes, carefully transfer undigested samples into freshly prepared collagenase type II solution and leave them for an additional hour of incubation.

5. Culture dish for chondron cultivation and their seeding density are selected according to the needs of downstream experiments.
6. For long term experiments, chondrons must be periodically (every second day) replaced to the new plates or flasks to get rid of adherent chondrocytes: they may interfere with experimental results. Maintaining chondrons on biomaterials such as alginate beads may contribute to longer preservation of the PCM [2].
7. Isolated chondrons can be used over the course of 5–7 days when replated every other day. Cell viability has to be determined prior to use (Fig. 5).

Acknowledgments

We wish to acknowledge funding from the European Structural and Social Funds through the Research Council of Lithuania (Lietuvos Mokslo Taryba) according to the Programme "Attracting Foreign Researchers for Research Implementation," Grant No 0.2.2-

LMTK-718-02-0022. Y.F.M.R. acknowledges financial support of the Dutch Scientific Research Council NWO/ZonMW VICI scheme (nr 91816631/528), the Leiden University Medical Center for supporting the ongoing RAAK study, and all RAAK study participants and members of the Meulenbelt lab at the LUMC.
Author disclosures: I.U., J.D., E.B., U.K., R.V., and Y.F.M.R. have no conflict of interest to disclose. AM has consulted for and received grants from the following companies during the last three years: Abbvie, Aché Laboratórios Farmacêuticos S.A., Sterifarma, Genacol, Galapagos, Kolon TissueGene, Pfizer Inc., Pfizer Consumer Health (PCH), GSK Consumer Healthcare, Servier, Bioiberica S.A., and Artialis S.A.

References

1. Mobasheri A et al (2019) The chondrocyte channelome: a narrative review. Joint Bone Spine 86(1):29–35
2. Vonk LA et al (2010) Preservation of the chondrocyte's pericellular matrix improves cell-induced cartilage formation. J Cell Biochem 110(1):260–271
3. Adam C et al (1998) The distribution of cartilage thickness in the knee-joints of old-aged individuals—measurement by A-mode ultrasound. Clin Biomech (Bristol, Avon) 13 (1):1–10
4. Gentili C, Cancedda R (2009) Cartilage and bone extracellular matrix. Curr Pharm Des 15 (12):1334–1348
5. Sophia Fox AJ et al (2009) The basic science of articular cartilage: structure, composition, and function. Sports Health 1(6):461–468
6. Bhosale AM, Richardson JB (2008) Articular cartilage: structure, injuries and review of management. Br Med Bull 87:77–95
7. Archer CW, Francis-West P (2003) The chondrocyte. Int J Biochem Cell Biol 35 (4):401–404
8. Manning WK, Bonner WM Jr (1967) Isolation and culture of chondrocytes from human adult articular cartilage. Arthritis Rheum 10 (3):235–239
9. Yonenaga K et al (2017) Optimal conditions of collagenase treatment for isolation of articular chondrocytes from aged human tissues. Regen Ther 6:9–14
10. Muhammad SA et al (2019) Optimization of protocol for isolation of chondrocytes from human articular cartilage. Cartilage. September 2019. doi:10.1177/1947603519876333
11. Guilak F et al (2018) Osteoarthritis as a disease of the cartilage pericellular matrix. Matrix Biol 71-72:40–50
12. Benninghoff A (1925) Form und Bau der Gelenkknorpel in ihren Beziehungen zur Funktion. Zweiter Teil: Der Aufbau des Gelenkknorpels in seinen Beziehungen zur Funktion. Zeitschr F Zellforsch U Mikroskop Anat 2(5):783–862
13. Poole CA et al (1987) Chondrons in cartilage: ultrastructural analysis of the pericellular microenvironment in adult human articular cartilages. J Orthop Res 5(4):509–522
14. Poole CA (1997) Articular cartilage chondrons: form, function and failure. J Anat 191(Pt 1):1–13
15. Zhang Z (2015) Chondrons and the pericellular matrix of chondrocytes. Tissue Eng Part B Rev 21(3):267–277
16. Wang QG et al (2008) Glycosaminoglycans in the pericellular matrix of chondrons and chondrocytes. J Anat 213(3):266–273
17. Lee GM et al (1997) Isolated chondrons: a viable alternative for studies of chondrocyte metabolism in vitro. Osteoarthr Cartil 5 (4):261–274
18. Sato T et al (2006) Comparative analysis of gene expression profiles in intact and damaged regions of human osteoarthritic cartilage. Arthritis Rheum 54(3):808–817
19. Ramos YF et al (2014) Genes Involved in the osteoarthritis process identified through genome wide expression analysis in articular cartilage; the RAAK Study. PLoS One 9(7): e103056

Chapter 3

Cartilage-Specific Cre Recombinase Transgenes/Alleles in the Mouse

Ioannis Kanakis, Mohammad Alhashmi, Ke Liu, Craig Keenan, Lorenzo Ramos Mucci, Blandine Poulet, and George Bou-Gharios

Abstract

Cartilage is a specialized skeletal tissue with a unique extracellular matrix elaborated by its resident cells, chondrocytes. The tissue presents in several forms, including growth plate and articular cartilage, wherein chondrocytes follow a differential differentiation program and have different fates. The induction of gene modifications in cartilage specifically relies on mouse transgenes and knockin alleles taking advantages of transcriptional elements primarily active in chondrocytes at a specific differentiation stage or in a specific cartilage type. These transgenes/alleles have been widely used to study the roles of specific genes in cartilage development, adult homeostasis, and pathology. As cartilage formation is critical for postnatal life, the inactivation or significant alteration of key cartilaginous genes is often neonatally lethal and therefore hampers postnatal studies. Gold standard approaches to induce postnatal chondrocyte-specific gene modifications include the Cre-loxP and Tet-ON/OFF systems. Selecting the appropriate promoter/enhancer sequences to drive Cre expression is of crucial importance and determines the specificity of conditional gain- or loss-of-function models. In this chapter, we discuss a series of transgenes and knockin alleles that have been developed for gene manipulation in cartilage and we compare their expression patterns and efficiencies.

Key words Chondrocyte, Cre-lox system, Transgenic mice, Enhancers, Gene expression

1 The Cre-loxP and Tet-ON/Tet-OFF Technologies

In vivo studies on gene functions and regulation and on the effects of specific mutations found in human diseases are often carried out using genetically modified mice. For this purpose, genes can be globally knocked out (inactivated) or knocked in (insertion of a desired mutation) using homologous recombination into mouse embryonic stem (ES) cells or fertilized oocytes. These gene alterations, however, can be early lethal or result in complex, multi-tissue phenotypes. To overcome such problems, conditional knockout (CKO)/knockin (CKI) technologies have been utilized. A main strategy consists in generating mice carrying

Tariq M. Haqqi and Véronique Lefebvre (eds.), *Chondrocytes: Methods and Protocols*, Methods in Molecular Biology, vol. 2245, https://doi.org/10.1007/978-1-0716-1119-7_3, © Springer Science+Business Media, LLC, part of Springer Nature 2021

both a tissue-specific Cre transgene and a floxed gene, such that the gene of interest is recombined only in cells that express Cre (Fig. 1). Cre is a 38-kDa tyrosine recombinase produced from the *Cre* (cyclization recombinase) gene of the bacteriophage P1 [1, 2]. It recognizes *loxP* (locus of X-over P1) sequences, that are palindromes of two 13-bp sequences flanking an 8-bp core (5' ATAACTTCGTATAATGTATGCTATACGAAGTTAT 3'; 34 bp in total). When two loxP sites are inserted in the same orientation, Cre recombines them into a single site and concomitantly deletes the intervening sequence (also known as loxP-flanked or floxed sequence) [3, 4]. When the two loxP sites are in opposite orientation, Cre induces inversion of the intervening sequence [5, 6].

Even CKO/CKI strategies can be early lethal or cause complex phenotypes. For example, the cartilage-specific *Col2a1* gene starts to be expressed in mid-gestation mouse embryos (E9.5, or E9.5) [7]. Therefore, Cre transgenes driven by *Col2a1* regulatory elements recombine floxed genes from this stage onwards. Thus, if genes are critical in early life, this strategy is not suitable to study

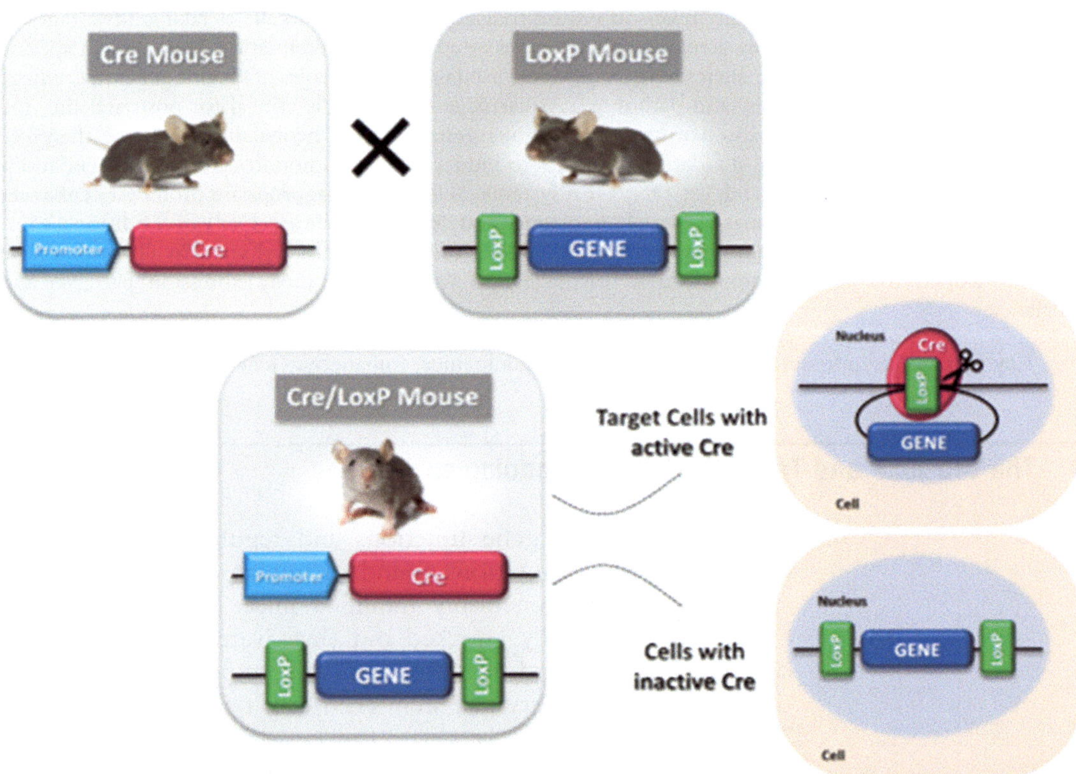

Fig. 1 Illustration of transgenic mouse generation using the conditional Cre-loxP gene manipulation. One transgenic mouse has a cell-specific regulatory sequence-driven Cre gene and the second transgenic mouse harbors a loxP-flanked (floxed) allele of the gene of interest. Expression of Cre recombinase results in recombination of the LoxP sites and thereby in inactivation (or other targeted modification) of the gene of interest

gene functions in adulthood. To circumvent this issue, inducible conditional systems (iCKO/iCKI) have been developed that permit modifying genes at a desired time. The two main inducible strategies use Cre-ERT [3, 5, 8] and the Tet-ON/Tet-OFF system [6, 9, 10].

Cre-ERT consists of a fusion of Cre with a mutated form of the ligand-binding domain of the estrogen receptor (ER). This mutant domain (ERT) binds tamoxifen, but not endogenous estrogen [11, 12]. Cre-ERT is sequestered in the cytoplasm by the heat shock protein 90 (Hsp90). Administration of tamoxifen (T) or 4-hydroxytamoxifen (4-OHT) causes disruption of the Hsp90/Cre-ERT complex, allowing Cre-ERT to localize to the nucleus, where recombination of loxP sites is achieved (Fig. 2). A newer version of Cre-ERT, Cre-ERT^2, is tenfold more sensitive to 4-OHT in vivo than Cre-ERT and is therefore preferred in most applications.

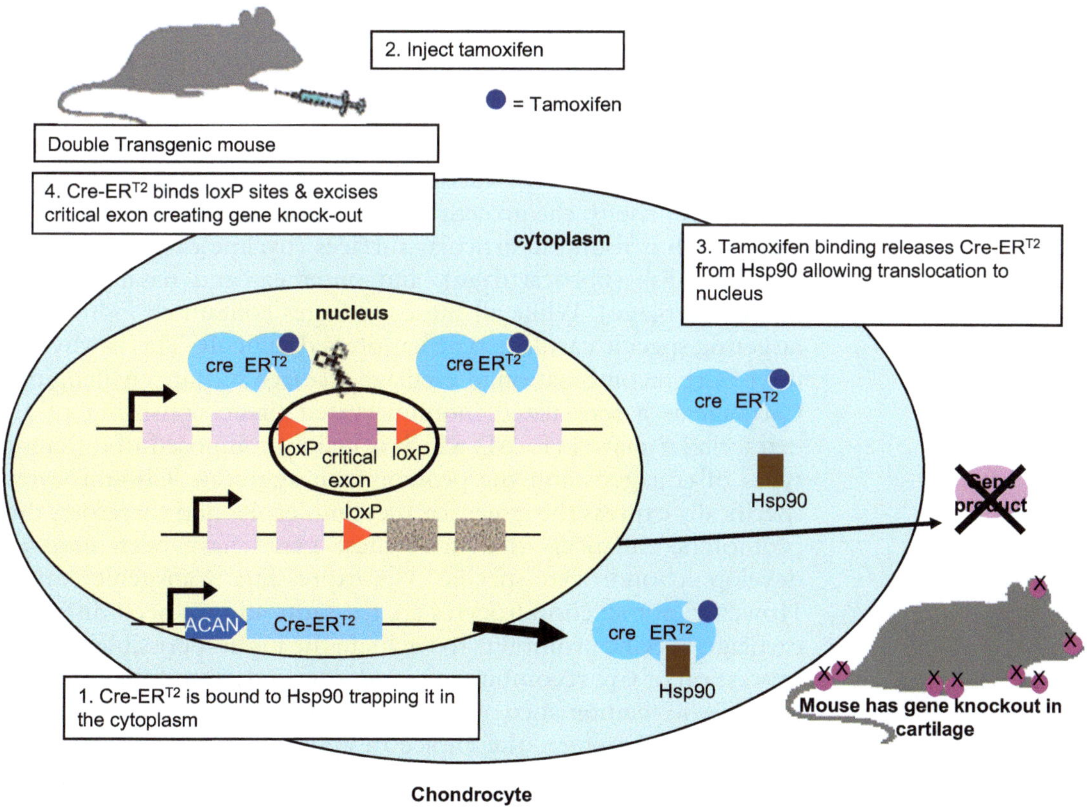

Fig. 2 Principle of the inducible Cre-loxP system using tamoxifen (Tam) induction of a modified estrogen receptor fused to Cre (CreER). In the absence of tamoxifen, CreER interacts with the heat shock protein 90 (HSP90) and is sequestered in the cytoplasm (1). Administration of Tam disrupts the interaction of HSP90 with CreER (2) which results in nuclear translocation of Cre (3). In the nucleus, CreER recognizes the loxP sites (4) and recombines the gene Y in tissue X (in this case chondrocytes as the Cre is driven by *Acan* regulatory sequences

An alternative system to achieve temporal and cell-specific inducibility is the tetracycline (Tet) system, which is often implemented using doxycycline (Dox), a tetracycline derivative. Dox is more efficient in regulating the Tet receptor (TetR) in comparison to tetracycline. This system can be applied in two modes, namely Tet-ON and Tet-OFF, for gene activation [13, 14] or inactivation [15], respectively. The Tet systems consist of three components: the reverse tetracycline-controlled transactivator (rtTA), the tetracycline-controlled transactivator (tTA) and the tetracycline-responsive element (TRE), also known as tetracycline operon (tetO), which controls gene expression, such as a Cre transgene. Activation of gene expression occurs after binding of rtTA to Dox and subsequent binding to $tetO_7$ (7 repeats of tetO) sequences (Tet-ON) (Fig. 3a). In the Tet-OFF mode, the normal binding of tTA to $tetO_7$ sequences is disrupted when tTA is coupled with Dox and therefore gene expression is inactivated (Fig. 3b). It is worth noting that, in most cases, unlike tamoxifen, doxycycline is administered via food or drinking water.

2 Chondrocyte-Specific Promoters

There are two types of cartilage, transient epiphyseal cartilage, which gives rise to the growth plate in bone, and permanent cartilage, which is found in articular surfaces (hyaline cartilage), intervertebral disk (fibrocartilage), and outer ear and nasal cartilage (elastic cartilage). While all these cartilages contain chondrocytes, targeting specific cartilage type has proved difficult. This is why the use of Cre recombination in cartilage research is quite challenging. Cartilage is a specialized, noninnervated tissue consisting of an extracellular matrix (ECM) which is mainly composed of different types of collagens and the proteoglycan aggrecan. Chondrocytes specifically express the genes for these molecules and therefore, the promoters/enhancers that drive these genes have been used to develop chondrocyte-specific Cre-expressing transgenic mice. However, not all chondrocytes are the same and we need different cartilage-specific promoters to facilitate the highest possible level of expression of Cre recombinase in order to understand the role that they play in maintenance of cartilage tissue and what goes wrong in pathology. Therefore, the choice of appropriate Cre drivers is of crucial importance since it can determine tissue-specific recombination. Below, we describe the main promoters/enhancers that have been used in cartilage biology (summarized in Table 1).

2.1 *Col2a1*

Collagen type II is an early chondrogenic marker and the *Col2a1* gene is expressed in all chondrogenic tissues of the axial and appendicular skeleton until the initiation of endochondral ossification. It is also transiently expressed in certain noncartilaginous tissues, such as notochord, eye, heart, epidermis, and brain [16]. The mouse

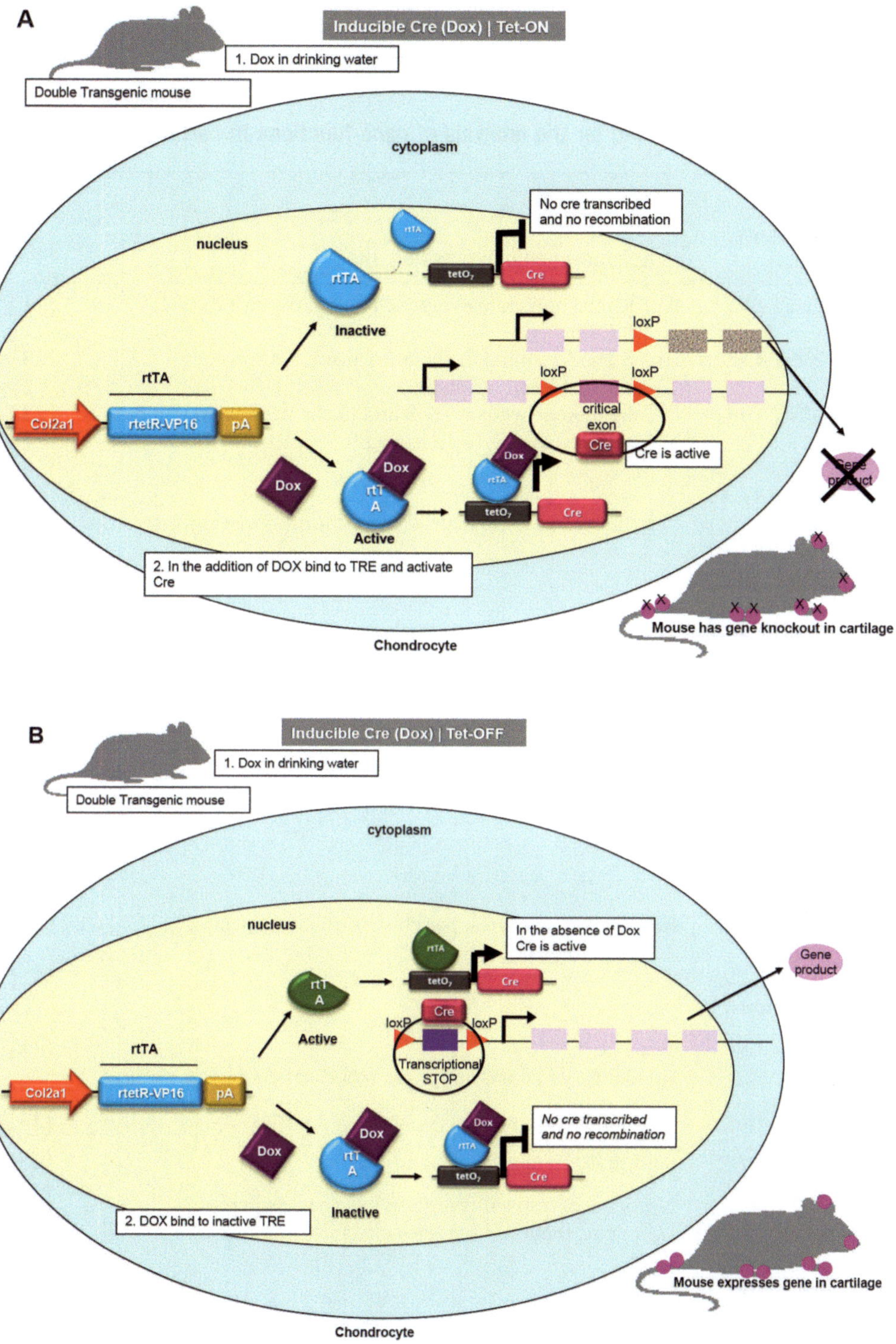

Fig. 3 Doxycycline (Dox)-induced tetracycline (Tet)-ON and -OFF systems. (**a**) In the Tet-ON system, ubiquitous or tissue-specific regulatory regions drive rtTA expression. In the absence of Dox, rtTA is unable to bind tetO7 (7 repeats of a 19-nucleotide-long tetO minimal enhancer, also referred to as TRE) Therefore, Cre is not expressed. Following Dox administration, Dox interacts with rtTA and activates it. Activated rtTA binds to tetO7 and thereby induces Cre expression in chondrocytes because the Cre is driven by *Col2a1* regulatory regions. (**b**) In the Tet-OFF system, in the absence of Dox, tTA is able to bind a tetO7 sequence (TRE) to induce Cre expression. Upon Dox administration, tTA is inactivated upon interaction with Dox and therefore Cre expression is prevented

Table 1
List of key *Cre* mouse lines used for the analysis of gene functions in cartilage

Genes	*Cre* transgene/alleles	Nonchondrogenic expression sites	References
Col2a1	*Col2a1-Cre*	Notochord and cranial mesenchyme between E8.75 and E9.0	[7]
	Col2a1-Cre	E11.5 LacZ staining in the notochord, sclerotome, neural tube and hindbrain.	[20]
	Col2a1-Cre	Cranial mesenchyme at E9.5 and in the neuroepithelium of the developing brain from E9.5 to E14.5	[22]
	Col2a1-CreERT		[23]
	Col2a1-CreERT²	Very limited in bone matrix (osteochondroprogenitors)	[24]
	Col2a1 DOX-Cre		[26]
	Col2a1-Cre10	Kidney epithelium in development E12.5	[54]
Acan/Agc1	*Agc1-AdeCre*		[31]
	Agc1-CreERT²		[32]
	Acan-CreERT²		[33, 34]
Col11a2	*Col11a2-Cre*	Notochord	[28]
Col10a1	*Col10a1-Cre*		[29]
	BAC-Col10-Cre		[30]
	Col10a1-Cre (10 kb)	Resting chondrocytes, perichondrium and fascia around digits	[31]
	Col10a1-Cre (enhancer)		[31]
Matn1	*Matn1-Cre*	Cruciate ligament, synovium, and some blood vessels	[41]
Gdf5	*Gdf5-Cre*	Tendon, brain, spinal cord, and hair follicles	[42]
Prg4	*Prg4-CreERT*	tendon, ligaments, heart, and liver	[43]
Sox9	*Sox9-Cre*	Notochord, testis, intestine, spinal cord, pancreas	[49]
	Sox9-CreERT²	Ligament, tendon, lung, kidney, brain, and heart	[50, 51, 63]
	BAC-Sox9-CreERT²	Pancreas	[52]

Col2a1 gene contains an enhancer in its first intron that is sufficient to drive the expression of its promoter specifically in cartilage in transgenic mice after E12.5 [17]. Lefebvre et al. [18, 19] showed that 309 bp of the promoter and 48 bp of the enhancer were sufficient for cartilage specificity, whereas shortening of the promoter to 89 bp and the enhancer to 18 bp was still sufficient for cartilage expression but also led to weak expression in brain and skin [19].

Several transgenic mouse lines have been generated using *Col2a1* promoter/enhancer segments to drive Cre expression with or without inducible capacity, such as Cre-ERT or Cre-ERT^2 systems. The first of these lines was created by Ovchinnikov et al. [7]. Its *Col2a1-Cre* transgene consisted of 3 kb of 5′ sequence upstream of transcription start site of the mouse gene, the first exon with a mutated initiation codon, 3 kb of the first intron, a splice acceptor sequence, an internal ribosome-entry site (IRES), Cre recombinase, and an SV40 large T antigen polyadenylation signal. Breeding this mouse line with a ROSA26-lacZ reporter strain (*R26R*) indicated that Cre-mediated recombination initiated in the notochord and cranial mesenchyme between E8.75 and E9.0, in somites at E9.5 and in the spine between E11.5 and E12.

The second *Col2a1-Cre* transgene was generated by Aszodi group who used an expression vector with *Col2a1* regulatory sequence containing a 3 kb region upstream of the transcription start site, exon 1 with the mutated initiation codon, 2.5-kb fragment of intron 1, a splice acceptor site followed by the nuclear localization signal of the SV40 large T antigen fused to the Cre recombinase coding sequences and finally exon 52 of the *Col2a1* gene containing the polyadenylation signal. This construct was injected into day one embryo [20]. Breeding this mouse line with an *R26R* reporter strain showed similar transgene expression in all chondrocytes as above but also expression in nonchondrogenic tissues (*see* Table 1). The third *Col2a1-Cre* line was generated by McMahon group using a modified *Col2a1* expression vector from [21] driving Cre recombinase and a *Col2a1* enhancer at the 3′ end. Three lines were described showing specific transgene staining in all cartilaginous structures at E12.5. Cre recombination occurred in all cartilage sites of the limbs from the stylopod to the autopod with variation in the intensity of the transgene between lines. All three lines also expressed the transgene in cranial mesenchyme and neuroepithelium of the developing brain [22]. Nakamura et al. [23] generated a *Col2a1-CreERT* transgenic mouse line, which included 1 kb of the mouse *Col2a1* proximal promoter, the ß-globin intron, SV40 polyA, and a 650-bp fragment of the *Col2a1* first intron enhancer. The authors conducted an extensive time course of expression following tamoxifen injection. They showed that Cre recombinase completed its job within 36 h after tamoxifen injection and transgene expression best matched that of the *Col2a1* mRNA at E12.5-E13 and beyond, in all cartilages (*see* **Note 1**).

In order to increase sensitivity to tamoxifen, Cre-ERT^2 was next utilized in association with the same *Col2a1* regulatory sequences as above [24]. Cre recombinase activity was reported in chondrocytes of the growth plate and articular cartilage embryos harvested at E18.5, that is, 3 days after tamoxifen injection. No expression was found in fibrocartilage of the intervertebral disk. What is notable about this line is that Cre recombinase activity was still efficient in articular cartilage when induced in mice beyond 3 months of age [25]. This result most likely reflected the integration site of this construct in the mouse genome.

The Tet-ON system was also used to generate a line of *Col2a1–Dox–Cre* transgenic mice, where Cre activity was detected in most cartilaginous tissues of E14.5 embryos via maternal doxycycline administration [26]. However, digits of the fore and hind limbs did not express the transgene. One-month-old mice demonstrated Cre recombinase activity in both growth plate and articular cartilage following 7 days of doxycycline treatment.

2.2 Col11a2

Collagen type XI, an important component of the cartilage collagenous network, is made of three subunits, which are encoded by *Col2a1*, *Col11a1*, and *Col11a2*. *Col11a2* is expressed only in cartilage [27] and was therefore a good candidate to generate Cre transgenes. Fujimaki et al. [28] generated transgenic mice expressing Cre under the control of a 742-bp regulatory sequence of mouse *Col11a2*. The authors detected Cre activity in all cartilaginous tissues from the early stage of chondrocyte differentiation at E12.5, and no expression in mesenchymal condensations. This transgene thus becomes active later in development than *Col2a1-Cre* transgenes.

2.3 Col10a1

The type X collagen gene (*Col10a1*) is specifically expressed in the hypertrophic chondrocytes of cartilage growth plates. *Col10a1-Cre* transgenes thus allow studying gene functions in these cells independently of functions in other chondrocytes. Gebhard et al. showed that a 4.6-kb mouse *Col10a1* regulatory sequence (−4410 to +634 bp), which includes a 500-bp enhancer, was sufficient to drive transgene expression in some but not all hypertrophic chondrocytes [29]. Aiming to generate a Cre transgene faithful to endogenous expression, they subsequently created transgenic mice using a bacterial artificial chromosome (BAC), in which they knocked in Cre recombinase into the *Col10a1* locus [30]. Cre activity was active as early as E13.5 in hypertrophic chondrocytes, in the vertebral growth plates of E16.5 and P1 mice, and also in subchondral bone marrow and endochondral bone trabeculae. More recently, Chen et al. [31] created two additional transgenic mouse lines. The first line used a 10-kb sequence, which contained the same distal enhancer as in the first line generated by Gebhard et al., up to exon 3. The second line was generated using four tandem copies of a 300-bp distal enhancer that was previously defined between −4296 and − 4147 upstream of the transcription start site of mouse *Col10a1*, the same region as the 500-bp enhancer described above. When crossed to *R26R*, the 10-kb line showed expression in resting chondrocytes, perichondrium as well as hypertrophic chondrocytes. In contrast, the 4x300-enhancer line showed expression exclusively in the hypertrophic zone of growth plates. The latter is possibly the best mouse line expressing Cre efficiently and specifically in hypertrophic chondrocytes.

2.4 Acan

Although aggrecan (*Acan*, formerly called *Agc1*) is the major proteoglycan in cartilage ECM, its regulation had not been extensively studied until recently. Henry et al. [32] generated a tamoxifen-inducible mouse line by knocking-in a *CreERT*^{2} cassette in the 3′-untranslated region of *Agc1*. Using an *R26R* reporter, these mice showed Cre activity in growth plate, hyaline cartilage, fibrocartilage of the menisci, trachea, and intervertebral disks. The most important finding in this study was the sustained expression of Cre in adult mice (*see* **Note 2**). As *Acan* mRNA has been reported in noncartilage tissues, investigators looked for cartilage specific *Acan* enhancers. Han and Lefebvre [33] identified a chondrocyte-specific enhancer approximately 10 kb upstream of the *Acan* transcription start site that was sufficient in transgenic mice to mimic the *Acan* expression pattern in embryonic and adult cartilage. Based on this observation, Lo Cascio et al. used this sequence with the minimal *Col2a1* promoter driving both Cre recombinase and the firefly luciferase reporter [34]. The expression was evidently stronger in hypertrophic cells of the ribs than any other chondrocytes and while most chondrocytes expressed the transgene in noncalcified articular cartilage, a few of them were positive in calcified articular cartilage [34].

Since this enhancer did not recapitulate the entire *Acan* expression pattern, we focused further on the *Acan* gene transcriptional regulation. Using ENCODE to delineate putative enhancers and testing them in transgenic reporter studies, we identified four new cis-acting elements active in chondrocytes [35]. In particular, one enhancer (−80 kb) distinguished itself from the others by being predominantly active in adult cartilage. Furthermore, the −62 kb element uniquely drove reporter activity in hypertrophic chondrocytes. The remaining chondrocyte-specific enhancers, +28 and −30 kb, showed no preference to chondrocyte type [35]. The transcription factor SOX9 interacted with all the enhancers in vitro and mutation of SOX9 binding sites in one of the enhancers (−30 kb) resulted in a loss of its chondrocyte specificity. Thus, we generated a new mouse line in which the −30 kb *Acan* enhancer was driving CreERT^2. However, instead of using a *Hsp68* minimal promoter, we used a 700-bp *Acan* promoter segment [36]. When tested with a Cre reporter, not all cartilage cells were positive for the transgene, unlike what we observed in every transgenic mouse line using this enhancer. This was not unique to one line, suggesting that the endogenous 700-bp proximal promoter may contain elements reducing the activity of this enhancer.

2.5 Matn1

Matrilins are noncollagenous proteins that are highly present in endochondral bone formation but weakly expressed in articular cartilage [37]. Among the four members of this family, matrilin-1 (*Matn1*) is known for binding to aggrecan [38] and collagen type II [39]. It is secreted in differentiated chondrocytes of hyaline cartilage [40] and its gene expression is maintained in cartilaginous tissues throughout life [37]. A mouse line has been developed in which the Cre recombinase gene was targeted to exon 1 of the *Matn1* gene [41]. *Matn1-Cre/R26R* mice showed Cre activity starting at E13.5. Importantly, this activity was limited to the cartilaginous anlagen of future bones, in epiphyseal but not in articular chondrocytes, which do not express *Matn1*. In comparison with *Col2a1-Cre* transgenes, this *Matn1-Cre* transgene was apparently not active in all chondrocytes, likely due to a low level of transgene expression [41].

2.6 Gdf5

Growth and differentiation factor 5 (*Gdf5*) is expressed in the presumptive joint mesenchyme of mouse embryos. Rountree et al. generated three *Gdf5-Cre* mouse lines using a 140-kb BAC containing the *Gdf5* locus modified by homologous recombination in bacteria to insert a Cre-internal ribosome entry site (IRES)-human placental alkaline phosphatase (hPLAP) cassette into the translation start site of *Gdf5* [42]. In *Gdf5Cre/R26R* embryos, transgene expression was specific to joints throughout development. Transgene expression was seen at E12.5 in the proximal joints, including the shoulder and knee. By E14.5, expression was seen in all but the most distal joints of the limbs. However, there was variability in both strength and extent of expression between different lines.

2.7 Prg4

Another promoter that has been recently used in cartilage biology is that of the *Prg4* gene, which encodes proteoglycan 4 (also called lubricin). Kozhemyakina et al. [43] generated knockin mice that express an inducible Cre recombinase driven by *Prg4* ($Prg4^{GFPCreERT2}$ mice) and administered tamoxifen at different ages. They found that *Prg4*-positive cells traced near the incipient joint surface at fetal stages serve as a progenitor population for all layers of mature articular cartilage. Furthermore, $Prg4^{GFPCreERT2}$ is primarily expressed by superficial chondrocytes in young mice, but expands into deeper zones of articular cartilage as mice age. *Prg4* is also expressed in other cells, including synovium-lining cells, tendon, ligaments, liver, and heart, as described in [44].

2.8 Sox9

Sox9 encodes a master regulator of chondrogenesis and is expressed in various progenitor cells such as chondroprogenitors, osteoprogenitors and preadipocytes [45]. In cartilage, *Sox9* remains highly expressed in differentiated chondrocytes of growth plates, but its expression stops when the cells become prehypertrophic

[46, 47]. *Sox9* is also expressed in adult healthy articular cartilage but is downregulated in osteoarthritic (OA) cartilage [48].

It is, therefore, not surprising that a number of studies have used *Sox9-Cre* to target chondrocytes in both growth plate and articular cartilage to unravel the role of various molecules in cartilage biology and OA pathophysiology. The first *Sox9-Cre* mouse was generated by de Crombrugghe laboratory, where a noninducible Cre recombinase was knocked in in the 3′ untranslated region of the *Sox9* gene [49]. This was followed by a tamoxifen inducible Cre ($CreERT^2$) inserted in the *Sox9* gene using the same method [50]. Lineage tracing with *R26R* concluded that *Sox9*-expressing precursors are the cellular origin of the cruciate ligament of the knee joint and the limb tendons [50]. In addition, using these mice, Ono et al. demonstrated that a subset of chondrogenic cells provides early mesenchymal progenitors in growing bones [51].

A second *Sox9-CreERT*^{2} mouse line was generated using a modified BAC clone RP23-229 L12 by inserting a KOZAK-*CreERT*^{2}-polyA sequence in place of the ATG start codon of the *Sox9* open-reading frame in exon 1. These mice were deposited at the Jax lab and used mainly in nonchondrogenic cells, namely *Sox9*$^+$ ductal cells of the adult pancreas [52]. Although these mice were mated to *R26R*, it is not known if they expressed the transgene in chondrocytes. While multiple enhancers active in chondrocytes have been identified in the *Sox9* locus, none has yet been found that is active exclusively in chondrocytes. The identification of such an enhancer would provide a great tool for specific targeting of cartilage.

3 Recommendation for the Use of the Cre-lox System to Target Chondrocytes

It is advisable that when a *Cre* mouse line is obtained from a laboratory or depository, the first experiment to be conducted should be to cross it with a reporter mouse such as *R26R* to ensure that the expression in your laboratory mirror what has been reported elsewhere, especially if you are targeting specific chondrocyte types (Fig. 4). Secondly, if an inducible system is preferred, we would recommend that the amount and the frequency of tamoxifen or Doxycycline to be used in mice be verified to efficiently recombine your gene of interest, as described in [23].

In breeding double transgenic mice, it is advisable to use *Cre* lines that are heterozygotes to enable the generation of *Cre*-negative mice, which then act as controls. Finally, when mating brothers and sisters from *Cre* transgenic lines where the transgene is integrated randomly in the genome, we found that *Cre* expression was reduced over time. It is therefore recommended that the *Cre* lines be crossed with wild type mice every few generations to maintain robust transgene activity.

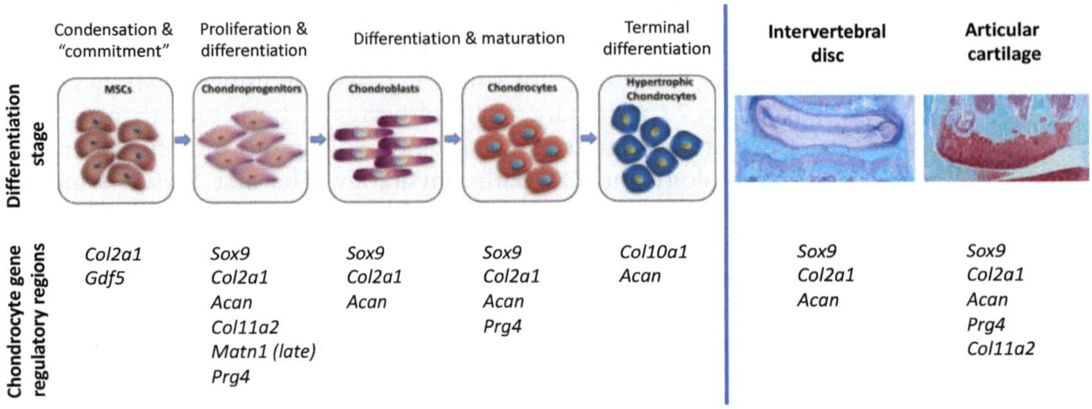

Fig. 4 Chondrocyte differentiation in early development, cartilage growth plates, intervertebral disks, and adult articular cartilage is accompanied by the activation of specific genes. The regulatory sequences of these genes have been used to generate various types of Cre transgenes and thereby provide routes to specifically target chondrocytes at desired differentiation stages

4 Concluding Remarks

The Cre-loxP system has come a long way from its first use in genetic manipulation. It has enabled us to pursue lineage-tracing experiments that have opened a new era of identifying chondrocyte stem cells, such as those reported recently in growth plate [53] and articular cartilage [43]. More importantly, it has changed our dogma about endochondral bone formation and the escape of hypertrophic chondrocytes from inevitable death to form bone [54]. The implication of such events offered an explanation why changes taking place in proliferating chondrocytes of the growth plates translate into bone phenotypes [55]. While we focused here on existing *Cre* mice that are cartilage-specific, there are other genes that have the potential of having chondrocyte-expressing cis-acting elements, such as *Ccn2* [56] and *Col6a1* [57]. Further, mesenchymal cells that are precursors for chondrocytes have been used to study limb development using a *Prrx1* promoter [58]. As genome manipulations become easier with CRISPR technology, we expect more refined systems to be developed in the future that may enable us to stratify chondrocytes and provide a new nomenclature with functional attributes that can stratify chondrocytes using the Cre-loxP spatiotemporal system.

5 Notes

1. Differences in activities and specificities have been reported among the various *Col2a1*-driven transgenes that have been generated by various groups. A major reason is likely that transgenes randomly integrate into the genome. Their activities

can thus be influenced positively, negatively, or ectopically by the local environment. For example, the cells targeted by the *Col2a1-CreERT* [23] and *Col2a1-Cre* [22] transgenes largely but not entirely overlap [59]. Furthermore, even though the *Col2a1* promoter and intron-1 enhancer drive gene expression primarily in cartilage, phenotypes in other tissues have been reported in some experiments. For example, conditional ablation of *Kif3a* and *Pkd1*, which encode primary cilium/intraflagellar transport-associated proteins using a *Col2a1-Cre* transgenic strain resulted in a severe form of polycystic kidney disease [60]. This was explained by transient expression of the *Col2a1-Cre* transgene during embryonic kidney morphogenesis. Furthermore, when this particular *Col2a1-Cre* mouse line [20] was tested in inflammatory joint disease, it was found that synovial fibroblasts in normal and inflamed synovium were *lacZ* positive, but *Cre* negative, suggesting that Cre-mediated recombination in joint interzone cells, which are the progenitors of both articular chondrocytes and synovial fibroblasts [61], endure in adult synovial cells despite the absence of ongoing *Cre* activity [62].

2. Recent studies by Rashid et al. 2017 have shown that the insertion of the CreERT2 cassette in Agc1 mouse which resulted in a deletion of 760 bp in the 3'UTR of the Acan gene, manifested in a cre dosage-dependent reduction in body weight and body length by one month of age due to a reduction in the length of the growth plate and the thickness of articular cartilage. Due to this effect on skeletal growth, heterozygous Agc1+/cre should be used for conditional deletion of a target gene in the cartilage tissue [64].

References

1. Nagy A (2000) Cre recombinase: the universal reagent for genome tailoring. Genesis 26 (2):99–109
2. Sternberg N, Hamilton D (1981) Bacteriophage P1 site-specific recombination. I. Recombination between loxP sites. J Mol Biol 150(4):467–486
3. Sauer B, Henderson N (1988) Site-specific DNA recombination in mammalian cells by the Cre recombinase of bacteriophage P1. Proc Natl Acad Sci U S A 85(14):5166–5170
4. Meinke G, Bohm A, Hauber J et al (2016) Cre recombinase and other tyrosine recombinases. Chem Rev 116(20):12785–12820
5. Branda CS, Dymecki SM (2004) Talking about a revolution: the impact of site-specific recombinases on genetic analyses in mice. Dev Cell 6 (1):7–28
6. McLellan MA, Rosenthal NA, Pinto AR (2017) Cre-loxP-mediated recombination: general principles and experimental considerations. Curr Protoc Mouse Biol 7(1):1–12
7. Ovchinnikov DA, Deng JM, Ogunrinu G et al (2000) Col2a1-directed expression of Cre recombinase in differentiating chondrocytes in transgenic mice. Genesis 26(2):145–146
8. Sauer B (1998) Inducible gene targeting in mice using the Cre/lox system. Methods 14 (4):381–392
9. Gu H, Zou YR, Rajewsky K (1993) Independent control of immunoglobulin switch recombination at individual switch regions evidenced through Cre-loxP-mediated gene targeting. Cell 73(6):1155–1164

10. Rajewsky K, Gu H, Kuhn R et al (1996) Conditional gene targeting. J Clin Invest 98 (3):600–603
11. Metzger D, Chambon P (2001) Site- and time-specific gene targeting in the mouse. Methods 24(1):71–80
12. Brocard J, Warot X, Wendling O et al (1997) Spatio-temporally controlled site-specific somatic mutagenesis in the mouse. Proc Natl Acad Sci U S A 94(26):14559–14563
13. Gossen M, Freundlieb S, Bender G et al (1995) Transcriptional activation by tetracyclines in mammalian cells. Science 268 (5218):1766–1769
14. Kistner A, Gossen M, Zimmermann F et al (1996) Doxycycline-mediated quantitative and tissue-specific control of gene expression in transgenic mice. Proc Natl Acad Sci U S A 93 (20):10933–10938
15. Gossen M, Bujard H (1992) Tight control of gene expression in mammalian cells by tetracycline-responsive promoters. Proc Natl Acad Sci U S A 89(12):5547–5551
16. Cheah KS, Lau ET, Au PK et al (1991) Expression of the mouse alpha 1(II) collagen gene is not restricted to cartilage during development. Development 111(4):945–953
17. Cheah KS, Au PK, Lau ET et al (1991) The mouse Col2a-1 gene is highly conserved and is linked to Int-1 on chromosome 15. Mamm Genome 1(3):171–183
18. Lefebvre V, Mukhopadhyay K, Zhou G et al (1996) A 47-bp sequence of the first intron of the mouse pro alpha 1(II) collagen gene is sufficient to direct chondrocyte expression. Ann N Y Acad Sci 785:284–287
19. Lefebvre V, Zhou G, Mukhopadhyay K et al (1996) An 18-base-pair sequence in the mouse proalpha1(II) collagen gene is sufficient for expression in cartilage and binds nuclear proteins that are selectively expressed in chondrocytes. Mol Cell Biol 16(8):4512–4523
20. Sakai K, Hiripi L, Glumoff V et al (2001) Stage-and tissue-specific expression of a Col2a1-Cre fusion gene in transgenic mice. Matrix Biol 19(8):761–767
21. Horton W, Miyashita T, Kohno K et al (1987) Identification of a phenotype-specific enhancer in the first intron of the rat collagen II gene. Proc Natl Acad Sci U S A 84(24):8864–8868
22. Long F, Zhang XM, Karp S et al (2001) Genetic manipulation of hedgehog signaling in the endochondral skeleton reveals a direct role in the regulation of chondrocyte proliferation. Development 128(24):5099–5108
23. Nakamura E, Nguyen MT, Mackem S (2006) Kinetics of tamoxifen-regulated Cre activity in mice using a cartilage-specific CreER(T) to assay temporal activity windows along the proximodistal limb skeleton. Dev Dyn 235 (9):2603–2612
24. Chen M, Lichtler AC, Sheu TJ et al (2007) Generation of a transgenic mouse model with chondrocyte-specific and tamoxifen-inducible expression of Cre recombinase. Genesis 45 (1):44–50
25. Zhu M, Chen M, Lichtler AC et al (2008) Tamoxifen-inducible Cre-recombination in articular chondrocytes of adult Col2a1-CreER (T2) transgenic mice. Osteoarthr Cartil 16 (1):129–130
26. Grover J, Roughley PJ (2006) Generation of a transgenic mouse in which Cre recombinase is expressed under control of the type II collagen promoter and doxycycline administration. Matrix Biol 25(3):158–165
27. Sugimoto M, Kimura T, Tsumaki N et al (1998) Differential in situ expression of alpha2(XI) collagen mRNA isoforms in the developing mouse. Cell Tissue Res 292 (2):325–332
28. Fujimaki R, Hayashi K, Watanabe N et al (2005) Expression of Cre recombinase in the mouse developing chondrocytes driven by the mouse alpha2(XI) collagen promoter. J Bone Miner Metab 23(3):270–273
29. Gebhard S, Poschl E, Riemer S et al (2004) A highly conserved enhancer in mammalian type X collagen genes drives high levels of tissue-specific expression in hypertrophic cartilage in vitro and in vivo. Matrix Biol 23(5):309–322
30. Gebhard S, Hattori T, Bauer E et al (2008) Specific expression of Cre recombinase in hypertrophic cartilage under the control of a BAC-Col10a1 promoter. Matrix Biol 27 (8):693–699
31. Chen J, Chen F, Bian H et al (2019) Hypertrophic chondrocyte-specific Col10a1 controlling elements in Cre recombinase transgenic studies. Am J Transl Res 11 (10):6672–6679
32. Henry SP, Jang CW, Deng JM et al (2009) Generation of aggrecan-CreERT2 knockin mice for inducible Cre activity in adult cartilage. Genesis 47(12):805–814
33. Han Y, Lefebvre V (2008) L-Sox5 and Sox6 drive expression of the aggrecan gene in cartilage by securing binding of Sox9 to a far-upstream enhancer. Mol Cell Biol 28 (16):4999–5013
34. Lo Cascio L, Liu K, Nakamura H et al (2014) Generation of a mouse line harboring a Bi-transgene expressing luciferase and

tamoxifen-activatable creER(T2) recombinase in cartilage. Genesis 52(2):110–119

35. Li IMH, Liu K, Neal A et al (2018) Differential tissue specific, temporal and spatial expression patterns of the Aggrecan gene is modulated by independent enhancer elements. Sci Rep 8 (1):950
36. Valhmu WB, Palmer GD, Dobson J et al (1998) Regulatory activities of the 5′- and 3-′-untranslated regions and promoter of the human aggrecan gene. J Biol Chem 273 (11):6196–6202
37. Klatt AR, Paulsson M, Wagener R (2002) Expression of matrilins during maturation of mouse skeletal tissues. Matrix Biol 21 (3):289–296
38. Hauser N, Paulsson M, Heinegard D et al (1996) Interaction of cartilage matrix protein with aggrecan. Increased covalent cross-linking with tissue maturation. J Biol Chem 271 (50):32247–32252
39. Winterbottom N, Tondravi MM, Harrington TL et al (1992) Cartilage matrix protein is a component of the collagen fibril of cartilage. Dev Dyn 193(3):266–276
40. Karcagi I, Rauch T, Hiripi L et al (2004) Functional analysis of the regulatory regions of the matrilin-1 gene in transgenic mice reveals modular arrangement of tissue-specific control elements. Matrix Biol 22(8):605–618
41. Hyde G, Dover S, Aszodi A et al (2007) Lineage tracing using matrilin-1 gene expression reveals that articular chondrocytes exist as the joint interzone forms. Dev Biol 304 (2):825–833
42. Rountree RB, Schoor M, Chen H et al (2004) BMP receptor signaling is required for postnatal maintenance of articular cartilage. PLoS Biol 2(11):e355
43. Kozhemyakina E, Zhang M, Ionescu A et al (2015) Identification of a Prg4-expressing articular cartilage progenitor cell population in mice. Arthritis Rheumatol 67 (5):1261–1273
44. Ikegawa S, Sano M, Koshizuka Y et al (2000) Isolation, characterization and mapping of the mouse and human PRG4 (proteoglycan 4) genes. Cytogenet Cell Genet 90 (3–4):291–297
45. Wang Y, Sul HS (2009) Pref-1 regulates mesenchymal cell commitment and differentiation through Sox9. Cell Metab 9(3):287–302
46. Ng LJ, Wheatley S, Muscat GE et al (1997) SOX9 binds DNA, activates transcription, and coexpresses with type II collagen during chondrogenesis in the mouse. Dev Biol 183 (1):108–121
47. Zhao Q, Eberspaecher H, Lefebvre V et al (1997) Parallel expression of Sox9 and Col2a1 in cells undergoing chondrogenesis. Dev Dyn 209(4):377–386
48. Brew CJ, Clegg PD, Boot-Handford RP et al (2010) Gene expression in human chondrocytes in late osteoarthritis is changed in both fibrillated and intact cartilage without evidence of generalised chondrocyte hypertrophy. Ann Rheum Dis 69(1):234–240
49. Akiyama H, Kim JE, Nakashima K et al (2005) Osteo-chondroprogenitor cells are derived from Sox9 expressing precursors. Proc Natl Acad Sci U S A 102(41):14665–14670
50. Soeda T, Deng JM, de Crombrugghe B et al (2010) Sox9-expressing precursors are the cellular origin of the cruciate ligament of the knee joint and the limb tendons. Genesis 48 (11):635–644
51. Ono N, Ono W, Nagasawa T et al (2014) A subset of chondrogenic cells provides early mesenchymal progenitors in growing bones. Nat Cell Biol 16(12):1157–1167
52. Kopp JL, Dubois CL, Schaffer AE et al (2011) Sox9+ ductal cells are multipotent progenitors throughout development but do not produce new endocrine cells in the normal or injured adult pancreas. Development 138(4):653–665
53. Mizuhashi K, Ono W, Matsushita Y et al (2018) Resting zone of the growth plate houses a unique class of skeletal stem cells. Nature 563(7730):254–258
54. Yang L, Tsang KY, Tang HC et al (2014) Hypertrophic chondrocytes can become osteoblasts and osteocytes in endochondral bone formation. Proc Natl Acad Sci U S A 111 (33):12097–12102
55. Javaheri B, Caetano-Silva SP, Kanakis I et al (2018) The Chondro-osseous continuum: is it possible to unlock the potential assigned within? Front Bioeng Biotechnol 6:28
56. Frost SL, Liu K, Li IMH et al (2018) Multiple enhancer regions govern the transcription of CCN2 during embryonic development. J Cell Commun Signal 12(1):231–243
57. Braghetta P, Vitale P, Piccolo S et al (1997) Tissue-specific expression of promoter regions of the alpha1(VI) collagen gene in cell cultures and transgenic mice. Eur J Biochem 247 (1):200–208
58. Logan M, Martin JF, Nagy A et al (2002) Expression of Cre recombinase in the developing mouse limb bud driven by a Prxl enhancer. Genesis 33(2):77–80
59. Nagao M, Cheong CW, Olsen BR (2016) Col2-Cre and tamoxifen-inducible Col2-

CreER target different cell populations in the knee joint. Osteoarthr Cartil 24(1):188–191

60. Kolpakova-Hart E, Nicolae C, Zhou J et al (2008) Col2-Cre recombinase is co-expressed with endogenous type II collagen in embryonic renal epithelium and drives development of polycystic kidney disease following inactivation of ciliary genes. Matrix Biol 27(6):505–512
61. Koyama E, Shibukawa Y, Nagayama M et al (2008) A distinct cohort of progenitor cells participates in synovial joint and articular cartilage formation during mouse limb skeletogenesis. Dev Biol 316(1):62–73
62. Fosang AJ, Golub SB, East CJ et al (2013) Abundant LacZ activity in the absence of Cre expression in the normal and inflamed synovium of adult Col2a1-Cre; ROSA26RLacZ reporter mice. Osteoarthr Cartil 21 (2):401–404
63. Balani DH, Ono N, Kronenberg HM (2017) Parathyroid hormone regulates fates of murine osteoblast precursors in vivo. J Clin Invest 127 (9):3327–3338
64. Harunur R, Haiyan C, Mohammad H, Amjad J (2017) Dwarfism in homozygous Agc1-CreERT mice is associated with decreased expression of aggrecan. Genesis 55(10). https://doi.org/10.1002/dvg.23070

Chapter 4

Isolation of Mouse Growth Plate and Articular Chondrocytes for Primary Cultures

Abdul Haseeb and Véronique Lefebvre

Abstract

Cartilage is a connective tissue presenting in several forms that are all essential components of the vertebrate skeleton. Complementing in vivo models, cultures of its resident cells—chondrocytes—are important experimental models in mechanistic and preclinical studies relevant to skeletal development and adult homeostasis and to such human pathologies as chondrodysplasias and osteoarthritis. Both growth plate and articular chondrocytes produce pancartilaginous extracellular matrix components, but the two cell subtypes also have distinct phenotypic properties that account for different structural features, functions, and fates of their tissues. Based on study goals, primary chondrocyte cultures should therefore be established from either growth plate or articular cartilage. Here, we describe the methods used in our laboratory to isolate and culture growth plate and articular chondrocytes from neonatal and adult mice, respectively. Both methods involve manual and enzymatic procedures to clean cartilage samples from contaminating tissues and to release chondrocytes as single-cell suspensions from their cartilage matrix.

Key words Chondrocytes, Cartilage, Growth plate, Articular, Primary culture, In vitro model

1 Introduction

Cartilage is an essential tissue in vertebrates. It comes in several subtypes, the main ones being growth plate and articular. Growth plates ensure skeletal elongation during fetal and postnatal development and contribute structurally and functionally to endochondral ossification, the process whereby most bones form [1, 2]. Mutations in genes controlling growth plate cartilage development and function underlie various forms of growth disturbances (dwarfism and overgrowth), and chondrodysplasias (generalized or localized malformations of the endochondral skeleton), and can also impair the endochondral process involved in bone fracture repair [3, 4]. While growth plates are temporary developmental structures, articular cartilage is a permanent tissue that provides a lubricated surface and resilient cushion for friction-free, deformation-free and pain-free movement of bone extremities

Tariq M. Haqqi and Véronique Lefebvre (eds.), *Chondrocytes: Methods and Protocols*, Methods in Molecular Biology, vol. 2245, https://doi.org/10.1007/978-1-0716-1119-7_4,

facing each other in synovial joints. Various factors can affect articular cartilage development and adult homeostasis [5, 6]. They include gene mutations, altered chondrocyte metabolism, excessive mechanical loading, inflammation, and aging [7]. All lead to degenerative joint diseases, wherein articular cartilage is progressively degraded and irreversibly eroded. Osteoarthritis, the most prevalent of these diseases, affects a large subset of the adult and aging human population [8].

All cartilage subtypes are made of an abundant extracellular matrix produced and maintained by chondrocytes, the only cells present in the tissue [9]. This matrix is composed of a fibrillar network implicating the collagen types 2, 9, and 11 [10] and a highly hydrated gel of aggrecan, other proteoglycans and glycoproteins [11]. In addition, each cartilage subtype features specific components, such as collagen type 10 in the hypertrophic zone of growth plates and the proteoglycan lubricin in the superficial layers of articular cartilage [12]. The differential compositions of these cartilage matrices imply that each chondrocyte is defined by its ability to implement a cartilage subtype- and zone-specific program in addition to the pancartilaginous differentiation program. Further, each chondrocyte is also defined by its proliferative or growth-arrested state and by its anabolic and catabolic rate. For instance, growth plate chondrocytes actively proliferate in the columnar zone of the tissue before growth-arresting and undergoing terminal hypertrophic maturation, whereas adult articular chondrocytes are not proliferative and do not normally proceed to terminal maturation. Each chondrocyte type is thus subjected to different modes of regulation and hence needs to be studied accordingly.

Primary cultures of chondrocytes have been used for decades as handy and potent tools in research projects aiming to dissect molecular mechanisms underlying the differentiated, anabolic and catabolic activities of chondrocytes in development, physiology, and pathology [13, 14]. They have also been used to rapidly and cost-effectively carry out pharmacological studies to test candidate drugs for the treatment of cartilage diseases [15]. Primary chondrocytes have traditionally been favored over cell lines. One reason is the lack of fully faithful chondrocytic cell lines [16]. Another is the opportunity of studying cells directly isolated from diseased human individuals and from mouse models of diseases. However, there are also challenges in using primary chondrocyte cultures. Namely, the amount of original tissue is generally small, the proliferation rate of chondrocytes, even in growth plates, is typically low, and cultured chondrocytes have been known for decades to lose or otherwise alter the phenotype that they had in their natural environment in vivo [17–19]. It is thus important when establishing primary chondrocyte cultures to be mindful of the advantages and limitations of the model systems and to have plans to validate preliminary findings made in vitro using preclinical animal models and, as appropriate, human individuals.

Here we describe methods used in our laboratory to prepare primary chondrocytes from neonatal mouse growth plate cartilage and from adult mouse articular cartilage. These methods are improved versions of previously published ones [20–23]. Growth plate chondrocytes are isolated from the thoracic cage of neonatal mice, and articular chondrocytes from the femoral heads of adult mice. In both cases, the cartilage samples are cleared of contaminating soft tissue by a brief digestion with Pronase followed by thorough washes. Chondrocytes are then released by digesting the cartilage extracellular matrix with bacterial collagenase for several hours to overnight. The cell suspensions are then washed to remove the enzymes and tissue debris, and the cells are counted and plated in cell culture dishes.

2 Materials

- For growth plate chondrocytes: neonatal mice (0–7 days of age).
- For articular chondrocytes: adult mice (6–8 weeks of age).
- Culture medium: Dulbecco's Modified Eagle Medium (DMEM) containing 4.5 g/l D-glucose (25 mM), 0.6 g/l L-glutamine (4 mM), and 110 mg/l sodium pyruvate (1 mM).
- Culture medium supplements: fetal bovine serum (FBS), and 100× solution of penicillin (10^4 U/ml) and streptomycin (10 mg/ml).
- Phosphate buffered saline (PBS) without calcium chloride and magnesium chloride.
- 70% Ethanol.
- Pronase from *Streptomyces griseus* (Sigma-Aldrich).
- Collagenase D from *Clostridium histolyticum* (Sigma-Aldrich).
- Liberase™ (medium thermolysin concentration) Research Grade (Sigma-Aldrich).
- Vacuum-driven filtration devices to sterilize solutions (pore size of 0.22 μm).
- Cell strainers, nylon (pore size of 40 μm).
- Regular cell culture supplies: serological pipettes, pipette tips, cell culture petri dishes, 15- and 50-ml conical centrifugation tubes.
- Dissection tools: scissors, forceps and scalpels.

3 Methods

All procedures must be performed according to relevant institutional and other regulations for proper laboratory practices. Unless otherwise indicated, all steps are carried out at room temperature and using a sterile technique in a biosafety cabinet.

3.1 Isolation of Chondrocytes from Growth Plate Cartilage

3.1.1 Mouse and Tissue Source

Costal cartilage constitutes the largest source of growth plate chondrocytes in the neonatal mouse. One mouse yields two to three million cells. The present protocol is written for one litter of pups (seven to ten mice). An alternative source, to be considered according to study goals, is to use the appendicular long bones. A detailed protocol using this source was previously published [21]. It is described as yielding immature articular chondrocytes. However, since it uses the entire epiphyseal cartilage tissue, it actually yields growth plate chondrocytes contaminated with a small percentage of articular chondrocytes. The yield was reported to be about one million chondrocytes per mouse. Like preparations from long bones, growth plate chondrocyte preparations from thoracic cages include a small percentage of articular chondrocytes. The latter cells originate from the sternocostal joints. In both cases, mouse pups should be used in the first week of age. If older mice are used, the growth plate chondrocyte preparations are also contaminated with cells from the ossification centers that form in the core region of the ribs and in the epiphyses of limb long bones.

3.1.2 Solutions

1. *Complete culture medium.* Add 55 ml of FBS and 5.5 ml of 100× penicillin/streptomycin solution to 500 ml of DMEM to give final concentrations of 10% serum, 100 U/ml penicillin, and 100 μg/ml streptomycin. Store at 4 °C. Use within a month.
2. *Pronase solution* (*see* **Note 1**). Dissolve enzyme powder in plain DMEM to a final concentration of 2 mg/ml. Dilute the solution with one volume of complete culture medium. Sterilize using a 0.22-μm filtration device. Use fresh or store in aliquot portions at −20 °C.
3. *Collagenase D solution* (*see* **Note 2**). Dissolve collagenase powder at 6 mg/ml in plain DMEM. Dilute the solution with one volume of complete culture medium to give a final concentration of 3 mg/ml. Sterilize using a 0.22-μm filtration device. Use fresh or store in aliquot portions at −20 °C.

3.1.3 Dissection of the Thoracic Cage

1. Euthanize mouse neonates (0- to 7-day-old pups) by hypothermia (*see* **Note 3**) by placing them in a plastic bag and submerging the bag in a mixture of water and wet ice for 30 min. Ensure death by decapitation.

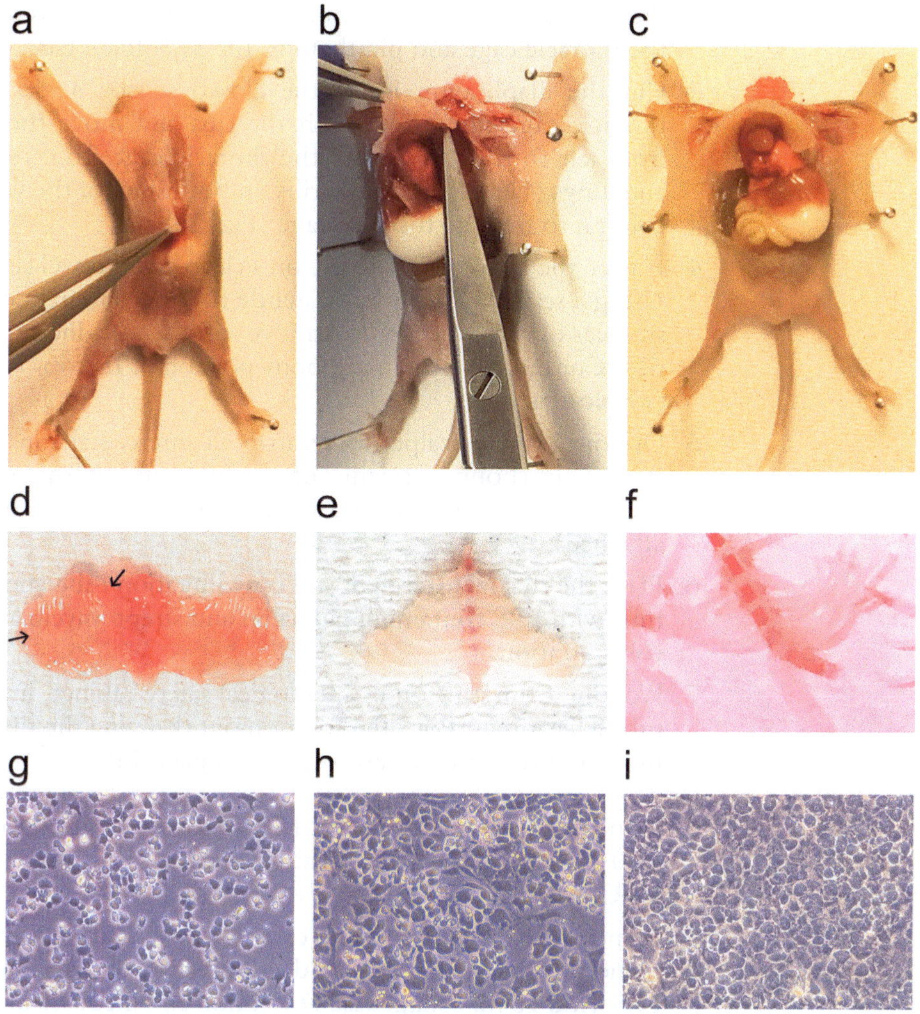

Fig. 1 Isolation and primary culture of growth plate chondrocytes from mouse neonate thoracic cages. (**a**) Euthanized mouse pup body in supine position, with legs stretched out and pinned to a dissection base. The thoracic cage is being exposed by incising the skin ventrally. (**b**, **c**) Dissection of the thoracic cage by cutting the ribs on the sides of the vertebral column. Note the skin flaps pinned to the dissection board. (**d**) Dissected thoracic cage before removal of soft tissues. Arrows point to the endochondral junction in the middle of the ribs. This junction is identified by the reddish color of bone. (**e**) Thoracic cage after the removal of the rib bony segments and the soft tissues covering the ribs. (**f**) Thoracic cage after the removal of intercostal muscle with Pronase and collagenase D digestion and PBS washes. Note that bony sternebrae (red) are still present. (**g**–**i**) Primary costal chondrocytes plated at a density of $10^5/cm^2$ and imaged under a phase-contrast microscope 1 day (**g**), 3 days (**h**) and 6 days (**i**) after plating

2. Working on a clean bench, place the mouse bodies in supine position on a Styrofoam base covered with a clean paper towel. Stretch the legs out and pin them to the base (Fig. 1a). Clean the skin with 70% ethanol. Make a ventral skin incision just below the thoracic cage, cut the skin along the sternum and

then along the lowest pair of ribs. Open the skin flaps to reveal the thoracic cage and pin them to the Styrofoam base.

3. Cut the diaphragm, insert the closed scissor blades in the thoracic cage just underneath the ribs and move the blades from left to right to isolate the thoracic cage from the heart and lungs (Fig. 1b). Cut the ribs on both sides of the vertebral column to free the thoracic cage (Fig. 1c).
4. Flatten the dissected structure on a clean paper towel (Fig. 1d). Using a scalpel or Kimwipes, gently rub the ventral and dorsal surfaces of the structure to remove soft tissues adhering to the ribs. Cut out any remaining bony portion of the ribs (Fig. 1e) (*see* **Note 4**).
5. Optional: use a scalpel to remove all intercostal muscle and sternebrae (bony segments of the sternum alternating with the cartilaginous intersternebrae to which the ribs are attached).
6. Place the ribs in a sterile 50-ml centrifugation tube and fill the tube with PBS containing penicillin and streptomycin at a 10× concentration.
7. Working from now on in a biological safety cabinet, invert the tube a few times. Pour the solution out (keeping the ribs in the tube) and repeat this wash twice with plain PBS.

3.1.4 Growth Plate Chondrocyte Release from Tissue and Plating

1. Incubate the thoracic cages in 15 ml Pronase solution for 1 h in a 10-cm petri dish in a cell culture incubator set at 37 °C and 5% CO_2 (*see* **Note 5**).
2. Remove the Pronase solution carefully by aspiration leaving the tissue behind in the culture dish. Wash the tissue once with PBS. Add 15 ml collagenase D solution and incubate in the cell culture incubator. Gently swirl the solution around for a few seconds every 10 min (or perform the incubation on a rocking platform placed in the incubator and set at low speed). This digestion step eliminates contaminating tissues and predigests the cartilage. It should not be extended beyond 30 min if the intercostal muscles were removed with scalpels. Otherwise, the digestion will take about 1 h for newborn mice and about 1 h 30 min for 7-day-old pups. After about 45 min of digestion, pipet the collagenase solution vigorously up and down a few times over the thoracic cages to force muscle dismantlement. Repeat this pipetting step every 10 min until all rib cartilage elements are clean.
3. Remove the soft tissue debris by taking advantage of the fact that they sediment slower than cartilage. Transfer the collagenase solution and ribs into a 50-ml conical centrifugation tube. As soon as the ribs reach the bottom of the tube, carefully pipet the collagenase solution out together with still-unsettled soft

tissue remnants (*see* **Note 6**). Fill the tube with PBS, invert it a few times, and remove the solution as soon as the ribs have sedimented. Repeat this procedure until all soft tissue contaminants are eliminated (Fig. 1f).

4. Digest the cleaned cartilage tissues for 4–5 h in 15 ml collagenase D solution in a petri dish in a 5% CO_2 incubator at 37 °C (*see* **Note 5**). Alternatively, dilute the collagenase solution three to four times in complete culture medium and let the digestion proceed overnight.
5. Release chondrocytes by pipetting the remaining cartilage fragments up and down a dozen times. Proceed gently to avoid damaging the cells.
6. Pass the digest through a 40-μm cell strainer over a 50-ml conical centrifugation tube to retain tissue remnants and obtain a single-cell suspension.
7. Wash the cells twice with 15 ml PBS, collecting the cells each time by centrifugation at 300 × *g* for 5 min.
8. Resuspend the cells in 10 ml complete culture medium. Count them using a hemocytometer or an automated cell counting device.
9. Plate the cells at a density of $10^5/cm^2$ in complete culture medium and place the dishes in a 37 °C/5% CO_2 incubator.
10. Most cells will attach to the plastic surface within a few hours to overnight (Fig. 1g). Dead cells will float and should be removed by a medium change after an overnight culture. Cell spreading and proliferation will be obvious within three days (Fig. 1h) and monolayers should reach confluence within five to seven days (Fig. 1i). Cartilage extracellular matrix produced de novo can be seen as a bright halo around rounding cells (*see* **Notes 7** and **8**).

3.2 Isolation of Chondrocytes from Articular Cartilage

3.2.1 Mouse and Tissue Source

Articular cartilage is still very immature in neonatal mice and is continuous with epiphyseal growth plate cartilage (Fig. 2a). Secondary ossification centers start to form when mice are in the second week of age. Articular cartilage then becomes progressively isolated from the epiphyseal growth plates, but it remains continuous with growth plate cartilage for several weeks (Fig. 2b). It acquires its adult cellular organization when mice reach 6–8 weeks of age (prepubertal to sexual maturity age). Its superficial half then remains unmineralized while its deeper half undergoes mineralization. Both halves contain only a few cell layers, and most cells are quiescent.

As mentioned earlier, the protocol published by Gosset and colleagues uses the entire cartilage ends of long bones from neonatal pups [21]. This tissue source thus yields growth plate chondrocytes mixed with a few immature articular chondrocytes. The

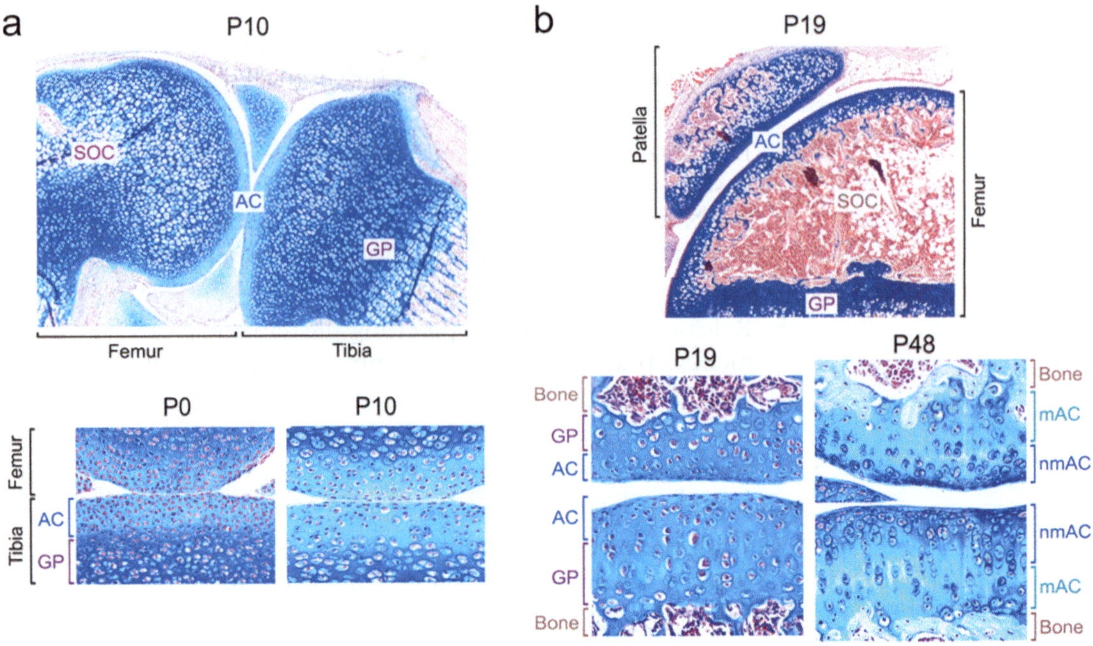

Fig. 2 Histological visualization of mouse articular cartilage maturation. (**a**) Pictures of sagittal sections through the knees of mice at postnatal day 0 (P0) and P10. Top, the femur and tibia epiphyses are largely cartilaginous (stained with Alcian blue), without any separation between articular (AC) and growth plate (GP) cartilage. By P10, secondary ossification centers (SOC) start to form in the middle of epiphyses. Bottom, high-magnification pictures of knee articular cartilage showing its continuity with growth plate cartilage at both P0 and P10. (**b**) Top, section through a mouse knee showing that secondary ossification centers have greatly expanded by P19 and now separate articular cartilage from epiphyseal growth plates. Some peripheral growth cartilage, however, remains continuous with articular cartilage, allowing further expansion of epiphyses. Bottom, high-magnification pictures showing that the knee articular cartilage is still continuous with growth plate cartilage at P19 but has acquired its adult organization by P48, with mineralized tissue (mAC) sandwiched between endochondral bone and nonmineralized cartilage (nmAC)

protocol published by Jonason and colleagues uses 3- to 4-week-old mice [23]. This protocol also yields a mixture of growth plate chondrocytes and immature articular chondrocytes.

The present protocol uses the femoral head cartilage caps of 6- to 8-week-old mice as a source of adult articular cartilage. These cartilaginous caps are thick and can be easily peeled off using blunt forceps. Other articular cartilage surfaces are not used because they are very thin and difficult to shave cleanly off underlying bone. One mouse brings in 10^4 to 10^5 chondrocytes. Although this yield is very low, it is mainly contributed by genuinely adult articular chondrocytes. As mice age, articular cartilage becomes less cellular and harder to separate from subchondral bone, yielding even fewer cells.

3.2.2 Solutions

See Subheading 3.1.2.

Liberase™ solution (*see* **Note 9**). Dissolve the content of the entire vial of lyophilized enzyme in plain DMEM to obtain a 5 mg/ml solution. Sterilize the solution using a 0.22-μm filtration device. Make aliquots and store at −20 °C. Prepare working solutions at 1 mg/ml in DMEM containing 1% serum.

3.2.3 Dissection of Articular Cartilage

1. Euthanize mice by CO_2 asphyxiation and ensure death by performing cervical dislocation.
2. Quickly dip the mice in 70% ethanol to wet and sterilize the fur. Remove excess ethanol with a paper towel.
3. Working on a clean bench, place the mouse bodies in supine position on a Styrofoam base covered with a clean towel and skin them from the waist down to the knees.
4. Using scalpels and forceps, dislocate the hip joint, keeping the femoral head intact (Fig. 3a).
5. Remove the cartilage cap of each femoral head using blunt forceps and place the tissues in PBS in a petri dish (Fig. 3b, c) (*see* **Note 10**).
6. Using scalpels, cut the cartilaginous caps into fragments of about 1 mm^3.
7. Place all cartilage pieces in a 50-ml centrifugation tube filled with PBS containing penicillin and streptomycin at a 10× concentration.
8. Work from now on in a biological safety cabinet, invert the tube a few times. Pour the solution out (keeping the cartilage pieces in the tube) and repeat this wash twice with plain PBS.

3.2.4 Articular Chondrocyte Release from Tissue and Plating

1. Incubate the cartilage fragments in 5 ml Pronase solution in a 60-mm cell culture dish for 1 h at 37 °C in a 5% CO_2 incubator (*see* **Note 5**). Gently swirl the solution around for a few seconds every 10 min (or use a rocking platform). This first step pre-digests the cartilage tissue.
2. Remove the Pronase solution with a manual pipette (*see* **Note 6**) and wash the cartilage fragments once with PBS.
3. Cover the cartilage pieces with 5 ml Liberase™ solution and place the dish back in the incubator for 6–8 h. This step disintegrates the cartilage matrix. It can be performed overnight, but the longer the digestion, the lower the yield of viable chondrocytes.
4. Liberate chondrocytes by pipetting the remaining cartilage fragments up and down a dozen times. Proceed gently to avoid damaging the cells.

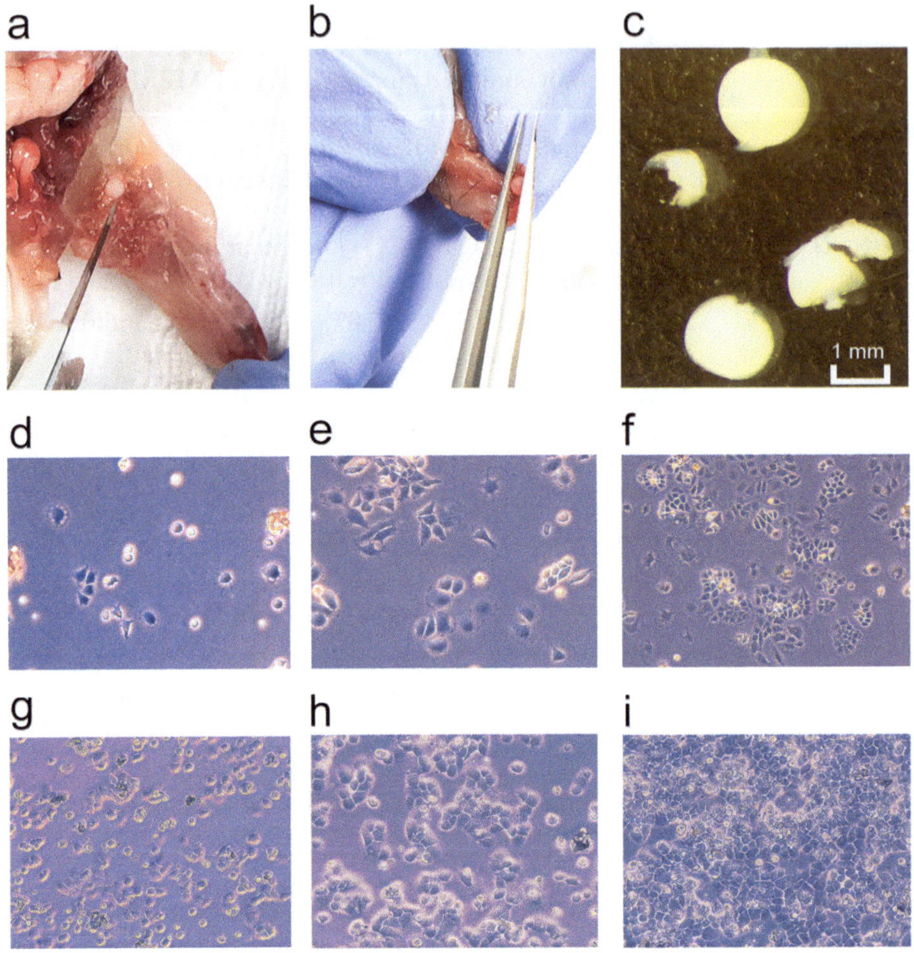

Fig. 3 Isolation and primary culture of adult mouse articular chondrocytes. (**a**) Picture showing a skinned mouse hind limb whose hip has been dislocated to expose the femoral head (white ball pointed at with a scalpel tip). (**b**) Snipping of the cartilage cap off a femoral head using blunt forceps. (**c**) Cartilage caps of femoral heads collected in a petri dish. (**d–f**) Pictures of primary chondrocytes 2 days (**d**), 4 days (**e**), and 6 days (**f**) after plating at low density. Note the formation of cell clones at days 4 and 8. (**g–i**) Pictures of primary chondrocytes 3 days (**g**) and 4 days (**h** and **i**) after plating at high density. The pictures in (**h**) and (**i**) were taken in the periphery and center of the dish, respectively

5. Pass the digest through a 40-μm cell strainer over a 50-ml conical centrifugation tube to retain tissue remnants and obtain a single-cell suspension.
6. Wash the cells twice with 15 ml PBS, collecting the cells each time by centrifugation at 300 × *g* for 5 min.
7. Resuspend the cells in 1 ml complete culture medium. Count them using a hemocytometer or an automated cell counting device.

8. Plate the cells at a density of 10^4 to $10^5/cm^2$ in complete culture medium and place the dishes in a 37 °C/5% CO_2 incubator.
9. Viable chondrocytes will attach to the plastic surface within 48 h. Dead cells will keep floating and should then be removed by a medium change. Cells plated at low density will form small colonies of closely connected polygonal cells within a week, but will hardly grow further (Fig. 3d–f). Cells plated at high density will form confluent monolayers of polygonal or round cells within a few days (Fig. 3g–i). Cartilage extracellular matrix produced de novo will be seen as a bright halo around rounding cells (*see* **Notes 7** and **8**).

4 Notes

1. Pronase is a mixture of extracellular proteolytic enzymes derived from the K-1 strain of *Streptomyces griseus*. It contains neutral protease, chymotrypsin, trypsin, carboxypeptidase, and aminopeptidase, as well as neutral and alkaline phosphatases. Pronase digests both native and denatured proteins.
2. Collagenase D is a mixture of proteolytic enzymes isolated from *Clostridium histolyticum*. It contains a lower tryptic activity and a higher collagenase activity than the Collagenase A and B preparations obtained from the same bacterium. We have obtained similar cartilage digestion rates using Collagenase B and Collagenase D. An alternative is to use Liberase™ Research Grade (*see* **Note 9**).
3. Hypothermia is used for euthanasia because mouse neonates are resistant to CO_2 asphyxiation and do not control their body temperature well (the opposite is true for older pups).
4. Ribs grow out and mature from the dorsal side (vertebral column) to the ventral side (sternum). In mouse neonates, the dorsal segment of the ribs is already ossified, whereas the ventral segment is still entirely cartilaginous.
5. All digestion steps are performed in cell culture dishes, not in centrifugation tubes. This is mandatory to keep cells alive.
6. Do not use a vacuum-connected pipette to avoid losing cartilage pieces through suction.
7. Primary chondrocytes have a high propensity to dedifferentiate in monolayer cultures. Experiments requiring bona fide chondrocytes should therefore be achieved with primary cultures a few days after plating. Plating the primary cultures at or near confluence and changing the medium every 3–4 days help maintain the differentiated cell phenotype. Low density favors

cell dedifferentiation. Frequent medium change (every day or other day) favors adipocytic differentiation.

8. Passaging of primary monolayers of chondrocytes cannot be achieved by incubation in trypsin–EDTA solution, but can be achieved by collagenase digestion for about 1 h. Passaging, however, is not recommended, because of cell dedifferentiation. If large numbers of cells are needed for experiments, it is best to start with more mice.
9. Liberase™ Research Grade is highly purified preparation of *Clostridium histolyticum* Collagenase I and Collagenase II that is formulated to contain the nonclostridial neutral protease Thermolysin at a Medium level (hence, TM). Liberase™ is recommended for applications that require a high yield of viable cells. We always use it for adult articular chondrocyte preparations.
10. The femoral head occasionally breaks while pulling out the cartilage cap. If this occurs, the cartilage can be shaved off from the subchondral bone using scalpels.

Acknowledgments

This work was supported by NIH grant AR072649 to V.L.

References

1. Olsen B, Reginato A, Wang W (2000) Bone development. Annu Rev Cell Dev Biol 16:191–220
2. Long F, Ornitz D (2013) Development of the endochondral skeleton. Cold Spring Harb Perspect Biol 5:a008334
3. Baron J, Savendahl L, De Luca F, Dauber A, Phillip M, Wit JM, Nilsson O (2015) Short and tall stature: a new paradigm emerges. Nat Rev Endocrinol 11(12):735–746. https://doi.org/10.1038/nrendo.2015.165
4. Warman ML, Cormier-Daire V, Hall C, Krakow D, Lachman R, LeMerrer M, Mortier G, Mundlos S, Nishimura G, Rimoin DL, Robertson S, Savarirayan R, Sillence D, Spranger J, Unger S, Zabel B, Superti-Furga A (2011) Nosology and classification of genetic skeletal disorders: 2010 revision. Am J Med Genet A 155A(5):943–968. https://doi.org/10.1002/ajmg.a.33909
5. Goldring MB, Marcu KB (2009) Cartilage homeostasis in health and rheumatic diseases. Arthritis Res Ther 11(3):224. https://doi.org/10.1186/ar2592
6. Houard X, Goldring MB, Berenbaum F (2013) Homeostatic mechanisms in articular cartilage and role of inflammation in osteoarthritis. Curr Rheumatol Rep 15(11):375. https://doi.org/10.1007/s11926-013-0375-6
7. Musumeci G, Aiello FC, Szychlinska MA, Di Rosa M, Castrogiovanni P, Mobasheri A (2015) Osteoarthritis in the XXIst century: risk factors and behaviours that influence disease onset and progression. Int J Mol Sci 16 (3):6093–6112. https://doi.org/10.3390/ijms16036093
8. Cross M, Smith E, Hoy D, Nolte S, Ackerman I, Fransen M, Bridgett L, Williams S, Guillemin F, Hill CL, Laslett LL, Jones G, Cicuttini F, Osborne R, Vos T, Buchbinder R, Woolf A, March L (2014) The global burden of hip and knee osteoarthritis: estimates from the global burden of disease 2010 study. Ann Rheum Dis 73 (7):1323–1330. https://doi.org/10.1136/annrheumdis-2013-204763
9. Archer CW, Francis-West P (2003) The chondrocyte. Int J Biochem Cell Biol 35

(4):401–404. https://doi.org/10.1016/s1357-2725(02)00301-1

10. Eyre DR, Weis MA, Wu JJ (2006) Articular cartilage collagen: an irreplaceable framework? Eur Cell Mater 12:57–63. https://doi.org/10.22203/ecm.v012a07
11. Roughley PJ, Lee ER (1994) Cartilage proteoglycans: structure and potential functions. Microsc Res Tech 28(5):385–397. https://doi.org/10.1002/jemt.1070280505
12. Rhee DK, Marcelino J, Baker M, Gong Y, Smits P, Lefebvre V, Jay GD, Stewart M, Wang H, Warman ML, Carpten JD (2005) The secreted glycoprotein lubricin protects cartilage surfaces and inhibits synovial cell overgrowth. J Clin Invest 115(3):622–631. https://doi.org/10.1172/JCI22263
13. O'Keefe RJ, Puzas JE, Loveys L, Hicks DG, Rosier RN (1994) Analysis of type II and type X collagen synthesis in cultured growth plate chondrocytes by in situ hybridization: rapid induction of type X collagen in culture. J Bone Miner Res 9(11):1713–1722. https://doi.org/10.1002/jbmr.5650091107
14. Thysen S, Luyten FP, Lories RJ (2015) Targets, models and challenges in osteoarthritis research. Dis Model Mech 8(1):17–30. https://doi.org/10.1242/dmm.016881
15. Zigon-Branc S, Jeras M, Blejec A, Barlic A (2017) Applicability of human osteoarthritic chondrocytes for in vitro efficacy testing of anti-TNFalpha drugs. Biologicals 45:96–101. https://doi.org/10.1016/j.biologicals.2016.09.013
16. Santoro A, Conde J, Scotece M, Abella V, Lopez V, Pino J, Gomez R, Gomez-Reino JJ, Gualillo O (2015) Choosing the right chondrocyte cell line: Focus on nitric oxide. J Orthop Res 33(12):1784–1788. https://doi.org/10.1002/jor.22954
17. Grundmann K, Zimmermann B, Barrach HJ, Merker HJ (1980) Behaviour of epiphyseal mouse chondrocyte populations in monolayer culture. Morphological and immunohistochemical studies. Virchows Arch A Pathol Anat Histol 389(2):167–187. https://doi.org/10.1007/bf00439484
18. Lefebvre V, Peeters-Joris C, Vaes G (1990) Production of collagens, collagenase and collagenase inhibitor during the dedifferentiation of articular chondrocytes by serial subcultures. Biochim Biophys Acta 1051(3):266–275. https://doi.org/10.1016/0167-4889(90)90132-w
19. Marijnissen WJ, van Osch GJ, Aigner J, Verwoerd-Verhoef HL, Verhaar JA (2000) Tissue-engineered cartilage using serially passaged articular chondrocytes. Chondrocytes in alginate, combined in vivo with a synthetic (E210) or biologic biodegradable carrier (DBM). Biomaterials 21(6):571–580. https://doi.org/10.1016/s0142-9612(99)00218-5
20. Lefebvre V, Garofalo S, Zhou G, Metsaranta M, Vuorio E, De Crombrugghe B (1994) Characterization of primary cultures of chondrocytes from type II collagen/beta-galactosidase transgenic mice. Matrix Biol 14(4):329–335. https://doi.org/10.1016/0945-053x(94)90199-6
21. Gosset M, Berenbaum F, Thirion S, Jacques C (2008) Primary culture and phenotyping of murine chondrocytes. Nat Protoc 3(8):1253–1260. https://doi.org/10.1038/nprot.2008.95
22. Mirando AJ, Dong Y, Kim J, Hilton MJ (2014) Isolation and culture of murine primary chondrocytes. Methods Mol Biol 1130:267–277. https://doi.org/10.1007/978-1-62703-989-5_20
23. Jonason JH, Hoak D, O'Keefe RJ (2015) Primary murine growth plate and articular chondrocyte isolation and cell culture. Methods Mol Biol 1226:11–18. https://doi.org/10.1007/978-1-4939-1619-1_2

Chapter 5

Fetal Growth Plate Cartilage: Histological and Immunohistochemical Techniques

Zachary Tata, Christophe Merceron, and Ernestina Schipani

Abstract

Skeletal development is a tightly regulated process that primarily occurs through two distinct mechanisms. In intramembranous ossification, mesenchymal progenitors condense and transdifferentiate directly into osteoblasts, giving rise to the flat bones of the skull. The majority of the skeleton develops through endochondral ossification, in which mesenchymal progenitors give rise to a cartilaginous template that is gradually replaced by bone. The study of these processes necessitates a suitable animal model, a requirement to which the mouse is admirably suited. Their rapid reproductive ability, developmental and physiologic similarity to humans, and easily manipulated genetics all contribute to their widespread use. Outlined here are the most common histological and immunohistochemical techniques utilized in our laboratory for the isolation and analysis of specimens from the developing murine skeleton.

Key words Chondrocyte, Cartilage, Growth plate, Development, Histology, Immunohistochemistry

1 Introduction

Bone development is an exquisitely intricate process requiring careful control, both spatially and temporally, by numerous molecular mechanisms and signaling pathways. Bone formation primarily occurs through two distinct processes. In the first, referred to as intramembranous ossification, mesenchymal condensations are formed during embryonic development. The cells of these condensations differentiate directly into osteoblasts and give rise to the flat bones of the skull. Alternatively, to form the majority of the skeleton, mesenchymal cells again condense during embryonic development, but differentiate to chondrocytes which form cartilaginous anlage, that is, the fetal growth plate [1–4]. This cartilage is then replaced by bone in a process known as endochondral ossification. Within the fetal growth plate, chondrocytes proliferate and deposit a matrix rich in type II, IX, and XI collagens, as well as specific proteoglycans, primarily aggrecan [5, 6]. These chondrocytes eventually form columnar structures, exit the cell cycle, and undergo

Tariq M. Haqqi and Véronique Lefebvre (eds.), *Chondrocytes: Methods and Protocols*, Methods in Molecular Biology, vol. 2245, https://doi.org/10.1007/978-1-0716-1119-7_5, © Springer Science+Business Media, LLC, part of Springer Nature 2021

hypertrophy, changing their matrix composition to one primarily enriched in type X collagen. Additionally, the hypertrophic chondrocytes express other markers such as matrix metalloproteinase thirteen (MMP-13) and vascular endothelial growth factor (VEGF), which are essential for the eventual degradation of the cartilaginous matrix and invasion with blood vessels, crucial steps for the replacement of the cartilaginous anlage with bone [7, 8]. Ultimately, terminally differentiated hypertrophic chondrocytes either die or transdifferentiate into cells of the osteoblastic lineage [9]. An overview of the E15.5 murine growth plate is provided in Fig. 1.

To further elucidate the processes driving skeletal development, the pathophysiological mechanisms behind diseases of these tissues, and eventually test novel treatment modalities, a suitable animal model is an indispensable asset. At present, the most widely used model for skeletal development is the mouse. Its small size and rapid reproductive ability make it an ideal tool for studying development, and mice share many developmental and physiological features with humans [10–13]. Furthermore, mice can be genetically manipulated with relative ease, and recent decades have been marked by numerous advances in the field made through the utilization of this model.

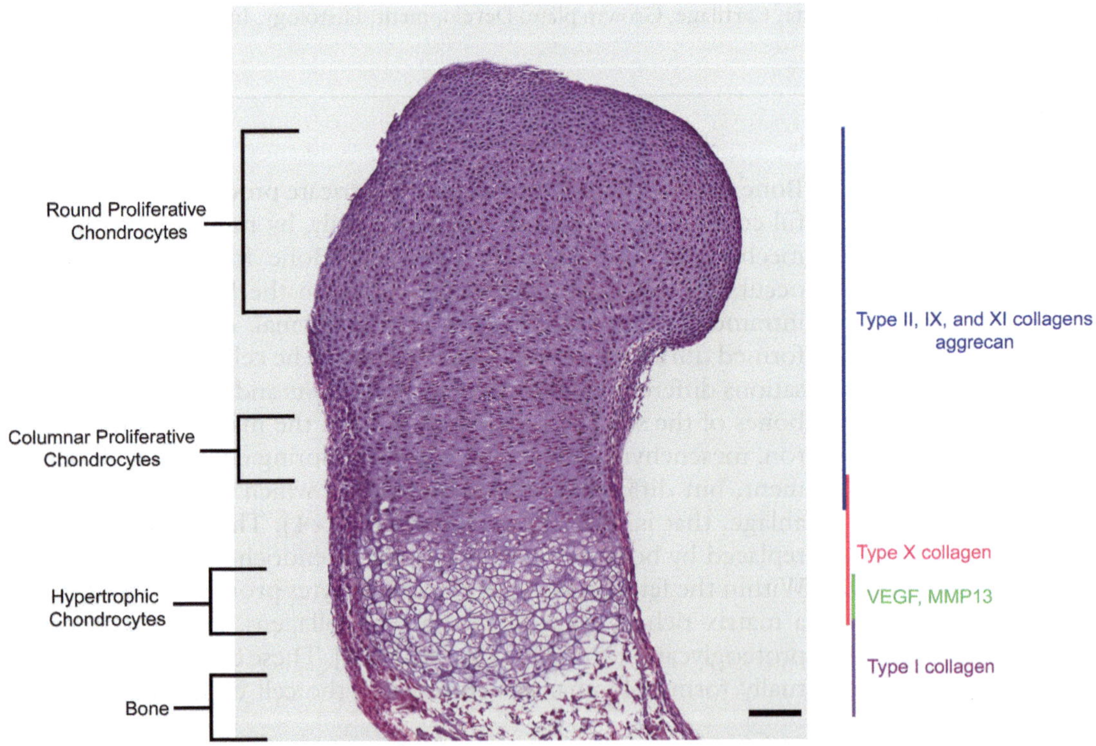

Fig. 1 Overview of an E15.5 murine proximal humerus growth plate histology following H&E staining. Scale bar = 100 μm

Described here are the most commonly utilized in vivo and in vitro techniques our laboratory currently uses for the analysis of the murine fetal growth plate during development. The first two protocols (*see* Subheading 2 for materials and Subheading 3 for methods) outline the procedure for collecting embryos and harvesting the necessary specimens for paraffin-embedded and fresh frozen tissues sections. Using these tissue sections, sections 3 and 4 demonstrate routine staining techniques: hematoxylin and eosin for morphological characterization and safranin O for growth plate glycosaminoglycan content analysis. Following this, section 5 describes the TUNEL assay, used for the in situ detection of cell death. Section 6 describes a standard immunohistochemical technique for the detection of antigens on paraffin-embedded tissues sections. Section 7 outlines the protocol for the EdU assay, which allows for the visualization of cellular proliferation. Section 8 describes EF5 staining, which is used to detect hypoxia in the developing growth plate through the use of immunofluorescence. Section 9 provides a method for the quantification of fluorescent signal in images using the ImageJ software.

2 Materials

2.1 Isolating Murine Embryos for Paraffin Sections

1. Pregnant mouse.
2. Isoflurane or CO_2 chamber.
3. Surgical instruments including scissors and forceps.
4. Ice.
5. 100 mm petri dishes.
6. 1× phosphate buffered saline (PBS).
7. 70% ethanol.
8. 20 mL scintillation vials.
9. Scissors.
10. 4% paraformaldehyde (PFA) in 1× PBS: Warm 540 mL distilled water until it reaches approximately 60 °C. Place in a 1 L glass beaker on a stir plate in a fume hood. Add 600 μL 10 N NaOH and 24 g paraformaldehyde powder. When the powder has dissolved, add 60 mL 10× PBS and turn off the heat. After it has cooled to room temperature, and 600 μL diethyl pyrocarbonate and stir for another 10 min. Using a pH strip, check that the solution is between 7.0 and 7.4. Store for up to 2–3 days at 4 °C, or aliquot and store up to 6 months at −20 °C.
11. Stereomicroscope.
12. Microdissection instruments.

13. Microsette biopsy cassettes.
14. Automated paraffin processor.
15. 80% ethanol.
16. 95% ethanol.
17. 100% ethanol.
18. Xylene.
19. Paraffin wax.
20. Embedding station.
21. Paraffin penetrated tissue samples in histology cassettes.
22. Appropriately sized molds.
23. Forceps for orienting specimens.
24. Scalpel.
25. Magnifying glass. Specifically, this is helpful for very small specimens.
26. Microtome.
27. Microtome blades.
28. Styrofoam cooler filled with ice and then water. This should be water surrounding a mass of ice, not ice floating in the water.
29. Forceps and brushes for manipulating sections.
30. Heated water bath.
31. Positively charged slides.
32. Paper towels.
33. Slide boxes.

2.2 Isolating Murine Embryos for Fresh Frozen Sections

1. Pregnant mouse.
2. Isoflurane or CO_2 chamber.
3. Surgical instruments including scissors and forceps.
4. Ice.
5. 100 mm petri dishes.
6. 1× phosphate buffered saline (PBS).
7. 70% ethanol.
8. Stereomicroscope.
9. Microdissection instruments.
10. Microsette biopsy cassettes.
11. Frozen section embedding media.
12. Plastic embedding molds.
13. Flat block of dry ice.
14. Forceps.

15. −80 °C freezer.
16. Positively charged slides.
17. Cryostat.
18. Cryostat sample holder.
19. High profile blades.
20. Paintbrushes and forceps for sectioning.
21. Slide box.

2.3 Fetal Growth Plate Hematoxylin and Eosin Staining

1. 5-μm paraffin-embedded tissue sections.
2. Slide warmer.
3. Slide rack.
4. Xylene in Coplin jars.
5. 100% ethanol in Coplin jars.
6. 95% ethanol in Coplin jars.
7. Distilled water in Coplin jars.
8. Harris Hematoxylin in Coplin jars.
9. Cold tap water.
10. Eosin Y in Coplin jars.
11. Xylene-based mounting media.
12. Glass coverslips.
13. Slide tray.

2.4 Safranin O Staining for the Evaluation of GAGs in the Murine Growth Plate

1. 5-μm paraffin-embedded tissue sections.
2. Slide warmer.
3. Slide rack.
4. Xylene in Coplin jars.
5. 100% ethanol in Coplin jars.
6. 95% ethanol in Coplin jars.
7. Distilled water in Coplin jars.
8. Paper filters.
9. Weigert's iron hematoxylin working solution: Mix 100 mL of solution A and 100 mL solution B directly before use. Working solution can be kept at room temperature for up to 24 h.
10. Running tap water.
11. 0.1% Fast Green working solution: Add 0.25 g Fast Green FCF to 250 mL distilled water. Cover and stir overnight. Filter directly before use.
12. 1% acetic water: Add 10 mL glacial acetic acid to 990 mL distilled water. May be made in advance.

13. 0.08% Safranin O working solution: Add 0.2 g Safranin O to 250 mL distilled water. Cover and stir overnight. Filter directly before use.
14. Xylene-based mounting media.
15. Glass coverslips.
16. Slide tray.

2.5 In Situ Cell Death Detection (TUNEL Assay)

1. 5-μm paraffin-embedded tissue sections.
2. Slide warmer.
3. 1× phosphate-buffered saline (PBS).
4. Permeabilization solution: Add 250 μL Triton X-100 and 0.25 g sodium citrate to 250 mL distilled water (chilled to 4 °C) in a cold room with stirring. Prepare fresh for every assay.
5. Humidified chamber.
6. Whatman paper.
7. Cold room.
8. Slide rack.
9. Xylene in Coplin jars.
10. 100% ethanol in Coplin jars.
11. 95% ethanol in Coplin jars.
12. PAP pen.
13. Roche In Situ Cell Death Detection Kit, Fluorescein Reaction Mixture: Combine 50 μL solution 1 and 450 μL solution 2.
14. 300 nM DAPI working solution: Make a 5 mg/mL stock solution by dissolving 10 mg of DAPI hydrochloride in 2 mL of distilled water. This can be divided into 50 μL aliquots and stored up to 2 years at −20 °C. Prepare the 300 nM working solution by diluting the stock solution 1:5000 in 1× phosphate buffered saline. This can be stored up to 6 weeks at 4 °C protected from light.
15. Prolong Gold Antifade Mountant.
16. Aluminum foil.
17. Slide box.
18. Fluorescent microscope.

2.6 Immunohistochemistry on Formalin-Fixed, Paraffin-Embedded Tissues

1. 5-μm paraffin-embedded tissue sections.
2. Slide warmer.
3. Slide humidification chambers.
4. 1× Tris buffered Saline (TBS).
5. Whatman paper.
6. Slide rack.

7. Xylene in Coplin jars.
8. 100% ethanol in Coplin jars.
9. 95% ethanol in Coplin jars.
10. Sodium citrate solution: Add 12.5 mL citrate buffer, pH 6.0, to 237.5 mL distilled water. Mix well and microwave the solution for approximately 2 min until it reaches 90–95 °C. Place in a heated water bath at 95 °C and allow solution to equilibrate for approximately 30 min. Prepare immediately before use.
11. Water bath set to 95 °C.
12. 3% hydrogen peroxide–methanol (H_2O_2–MeOH): Add 25 mL of 30% H_2O_2 to 225 mL methanol. Prepare immediately before use.
13. Distilled water in Coplin jar.
14. Laboratory wipes.
15. PAP pen.
16. TNB: Add 0.05 g NEN block reagent (from TSA kit) to 10 mL of 1× TBS while heating to dissolve the block reagent. Can be stored up to 2 years at −20 °C.
17. Unconjugated primary antibody.
18. 1× Tris–NaCl–Tween 20 (TNT): Add 400 mL of 10× TBS and 2 mL of Tween 20 to 3600 mL distilled water. Preparation is easiest if Tween 20 is added while stirring the solution.
19. Biotin-conjugated secondary antibody.
20. Rocking platform.
21. TSA kit. Available through Akoya Biosciences. The kit will contain biotinyl tyramide, amplification diluent, and HRP-SA.
22. DAB (3,3′-diaminobenzidine) Peroxidase (HRP) substrate kit Solution: Available from Vector Labs. Add two drops of the buffer solution to 5 mL distilled water and mix well by vortexing while protecting from light. Next add four drops of DAB stock solution and vortex to mix. Prepare immediately before use.
23. Mayer's hematoxylin.
24. 1× phosphate-buffered saline.
25. Cold running tap water.
26. Xylene-based mounting media.
27. Glass coverslips.
28. Slide tray.

2.7 EdU Assay for Evaluating Cellular Proliferation

1. Pregnant mouse.
2. Click-iT EdU Alexa Fluor 488 Imaging Kit.
3. Ice.
4. 1× phosphate-buffered saline (PBS).
5. 70% ethanol.
6. Balance.
7. 1 mL syringes equipped with 27-G needles.
8. 100 mm petri dishes, sterile.
9. Dissection instruments.
10. 5-μm paraffin-embedded tissue sections.
11. Slide warmer.
12. Humidified chamber.
13. Whatman paper.
14. Xylene in Coplin jars.
15. 100% ethanol in Coplin jars.
16. 95% ethanol in Coplin jars.
17. Click-iT EdU Alexa Fluor 488 Imaging Kit Reaction Mixture: Available through Invitrogen. Add reagents in the following order: 430 μL 1× Click-iT reaction buffer, 20 μL $CuSO_4$ (component E), 1.2 μL Alexa Fluor azide working solution, 50 μL 1× Click-iT EdU buffer additive working solution. Prepare immediately before use.
18. PAP pen.
19. Laboratory wipes.
20. Hoechst 33342 working solution: Dilute Hoechst 33342 (component G of the Click-iT kit) 1:2000 in 1× PBS in the dark. Prepare immediately before use and protect from light.
21. Prolong Gold Antifade Mountant.
22. Glass coverslips.
23. Aluminum foil.

2.8 EF5 Staining to Detect Hypoxia in the Developing Growth Plate

1. Balance.
2. Pregnant mouse.
3. 1 mL syringe.
4. 27 G needle.
5. 10 mM EF5 stock: Add 3.02 mg EF5 to 1 mL of 0.9% NaCl. Sonicate for 4 h at 37 °C until the powder is fully dissolved. May be stored up to 1 year at −20 °C.
6. Materials for cesarean section embryo isolation. Refer to fresh frozen isolation protocol (Subheading 2.2).

7. Ice.
8. Sterile 1× phosphate buffered saline (PBS).
9. 100 mm petri dish.
10. 70% ethanol.
11. Materials for dissection, processing, and sectioning fresh frozen specimens. Refer to fresh frozen isolation protocol (Subheading 2.2).
12. 10 μm freshly frozen tissue sections.
13. 100% acetone.
14. Humidified chamber.
15. Whatman paper.
16. Slide racks.
17. 250 mL Wheaton glass staining dishes.
18. Laboratory wipes.
19. PAP pen.
20. 5% mouse serum–PBS blocking solution: Add 0.5 mL normal mouse serum to 9.5 mL 1× PBS. Mix well and divide into 400 μL aliquots. Store up to 1 year at −20 °C.
21. 3% bovine serum albumin–PBS: Add 300 mg bovine serum albumin (BSA) to 10 mL of 1× phosphate buffered saline. Mix well until the powder is fully dissolved and store at 4 °C until use. Prepare directly before the staining.
22. Distilled water.
23. 300 nM DAPI working solution: Make a 5 mg/mL stock solution by dissolving 10 mg of DAPI hydrochloride in 2 mL of distilled water. This can be divided into 50 μL aliquots and stored up to 2 years at −20 °C. Prepare the 300 nM working solution by diluting the stock solution 1:5000 in 1× phosphate buffered saline. This can be stored up to 6 weeks at 4 °C protected from light.
24. Prolong Gold Antifade Mountant.
25. Glass coverslips.
26. Aluminum foil.
27. Fluorescent microscope.

2.9 Fluorescent Image Quantification Using ImageJ

1. DAPI Image for nuclei quantification or total cell number.
2. Fluorescent signal image for positive cell quantification.
3. ImageJ software.
4. Adobe Photoshop software.

3 Methods

3.1 Isolating Murine Embryos for Paraffin Sections

Endochondral bone development analysis can begin as early as E11.5, when the mesenchymal condensation begins to form. This protocol details how to isolate, fix, dissect, process, and section tissues, specifically the limbs, from murine embryos that have been embedded in paraffin. Briefly, after isolation of the embryos, they are fixed. Though dissection can occur before or after fixation, the fixative step preserves the tissue and gives the investigator more time to properly dissect the specimen. Following this, the tissues are processed using an automated processor. This consists of dehydration in a gradient of ethanols, clearing through the use of xylene, and infiltration with paraffin wax. The tissues are then embedded in a block of paraffin wax suitable for clamping into the chuck of a microtome during sectioning, and finally sectioned. These sections are useful for a number of staining and immunohistochemical assays.

1. Sacrifice the pregnant female with isoflurane or CO_2.
2. Spray the mouse thoroughly with 70% ethanol.
3. Pinch the abdominal skin and make a longitudinal incision, approximately 1 cm in length, with surgical scissors. Pull the skin apart to a sufficient distance to expose the abdominal cavity.
4. Using the scissors, carefully open the peritoneum and move the intestines to expose the uterus.
5. Remove the uterus by cutting the connective tissue and blood vessels, taking care to not damage any of the embryos.
6. Rinse the uterus in 1× PBS in a 100 mm petri dish and immediately move to a fresh 100 mm petri dish filled with 1× PBS on ice.
7. Peel the soft tissue with sterile forceps and quickly remove each embryo from its amniotic sac.
8. Place the embryos in 1× PBS in a clean 100 mm petri dish.
9. Rinse each embryo 3 times in sterile 1× PBS, making sure to clean your instruments with 70% ethanol in between handling each embryo.
10. Fill 20 mL scintillation vials with 4% paraformaldehyde (PFA) in 1× PBS.
11. Rinse each embryo an additional time in 1× PBS.
12. Using scissors, open each embryo's abdomen to facilitate rapid entrance of the fixative. Take care not to damage any of the organs or ribs.
13. Place 1 embryo in each filled scintillation vial.

14. Store the vials for 48 h at 4 °C, replacing the 4% paraformaldehyde (PFA) in 1× PBS with fresh 4% paraformaldehyde (PFA) in 1× PBS after 24 h.
15. After 48 h have elapsed, remove the 4% paraformaldehyde (PFA) in 1× PBS and replace with 70% ethanol. Specimens can be stored at 4 °C in ethanol until dissection (*see* **Notes 1** and **2**).
16. Place a clean 100 mm petri dish with 5–10 mL of 1× PBS on the stage of a stereomicroscope.
17. Remove the fixed embryo from the scintillation vial and place it in the petri dish.
18. Select the forelimb or hind limb of interest, and remove it by cutting the proximal end with microdissecting scissors.
19. Remove the skin from the limb carefully with microdissecting forceps.
20. Carefully remove the soft tissues surrounding the bone or cartilage, take care not to damage the growth plate or bone.
21. Place the dissected specimen in a Microsette biopsy cassette, and store in 70% ethanol at 4 °C until processing can be completed.
22. Place the cassettes into the automated processor (*see* **Note 3**).
23. Add 70% ethanol for 1 h (*see* **Note 4**).
24. Add 80% ethanol for 1 h.
25. Add 95% ethanol for 1 h.
26. Add 95% ethanol for 1 h.
27. Add 100% ethanol for 1 h.
28. Add 100% ethanol for 1 h.
29. Add xylene for 1 h (*see* **Note 5**).
30. Add xylene for 1 h.
31. Add paraffin wax at 58 °C for 1 h under vacuum (*see* **Note 6**).
32. Add paraffin wax at 58 °C for 1 h under vacuum.
33. Add paraffin wax at 58 °C for 1 h under vacuum.
34. Remove cassettes following the completion of the program.
35. Allow cassettes to equilibrate in the paraffin reservoir of the embedding center for approximately 20 min prior to embedding.
36. Fill an appropriately sized mold with paraffin.
37. Using a magnifying glass if necessary, orient the specimens as described in the Subheading 4 of this protocol (*see* **Note 7**).
38. While holding the specimens carefully in place with forceps, gently move the mold over the cold plate.

39. Hold the specimen in position until the wax on the bottom of the mold freezes, securing the specimen in place.
40. Place the top of the cassette over the mold, and fill completely with paraffin.
41. Return the mold to the cold plate to allow the block to freeze completely, approximately 20 min.
42. Gently remove the mold from the block and trim excess paraffin away from the block using a scalpel (*see* **Note 8**).
43. Set a histology water bath to approximately 55 °C.
44. Orient the blade on the microtome to approximately 5°.
45. Set the microtome to trimming depth (approximately 10 μm) and section away excess paraffin until the tissue is just exposed in the block.
46. Place the block in the ice water bath, with the tissue section submerged for 5–10 min (*see* **Note 9**).
47. Set the microtome cutting depth to 5 μm.
48. Place the cooled block in the microtome chuck.
49. Orient the block for optimal cutting (*see* **Notes 10** and **11**).
50. Begin sectioning, carefully pulling the ribbon away from the cutting surface.
51. Gently detach the ribbon with a fine paintbrush.
52. Float the ribbon on the surface of the water bath.
53. Collect sections by dipping the slides into the water and gently bringing them to the surface underneath the floating sections.
54. Remove excess water by tapping the slides on a paper towel.
55. Place the slides into a slide box.
56. Store the slides up to 1 year at 4 °C.

3.2 Isolating Murine Embryos for Fresh Frozen Sections

Frozen sections have inherent disadvantages when compared to their paraffin counterparts, such as substandard morphology and diminished resolution at higher magnifications. Additionally, they are more technically demanding to cut. However, the use of frozen sections is, nonetheless, essential for the demonstration of many antigens for use in immunohistochemistry, sensitive detection of minimally expressed mRNAs, and temperature-sensitive enzymatic reactions, such as alkaline phosphatase. The following protocol outlines the procedures necessary to obtain quality freshly frozen sections.

1. Isolate embryos as described for paraffin sections (Subheading 3.1).
2. Dissect embryos immediately after isolation.

3. Label plastic base molds and partially fill with embedding medium.
4. Place the tissue into the base mold and orient.
5. Flatten the tissue against the bottom of the mold with gentle pressure from a pair of forceps.
6. Let the specimen equilibrate in the media for 1–2 min at room temperature.
7. Place the mold on a flat block of dry ice and hold the tissue flat against the bottom of the mold until the media freezes sufficiently to hold the tissue in place (*see* **Note 12**).
8. Allow the media to continue freezing until it is entirely solid, approximately 10 min.
9. Store the block at −80 °C immediately after freezing until use (*see* **Note 13**).
10. Equilibrate the frozen blocks, sectioning instruments, and sample holder in the cryostat for approximately 15 min at −20 °C.
11. Apply approximately 0.5 mL of embedding media to the sample holder.
12. Place the frozen block in the sample holder and allow it to freeze firmly into place.
13. Orient the blade of the cryostat to approximately 30°.
14. Select a 10 μm section thickness.
15. Place the sample holder in the cryostat chuck and orient the block for optimal cutting (*see* **Notes 14** and **15**).
16. Using the forceps and paintbrush, pick up one or two sections per slide.
17. Store the slides in a slide box at −80 °C for long term storage, or −20 °C for shorter periods (*see* **Note 16**).

3.3 Fetal Growth Plate Hematoxylin and Eosin Staining

Paraffin-embedded tissue sections are suitable for a broad array of stains. Hematoxylin and eosin (H&E) is an essential histological staining, and allows for the evaluation of tissue morphology, including the developing growth plate [14–18]. Hematoxylin, a basic/positive stain, binds to acidic and negatively charge substances, such as nucleic acids, and generates a violet/blue nuclear staining. Eosin is acidic/negative and preferentially binds positively charged substances such as proteins, resulting in a pink cytoplasmic staining. While there are a variety of kinds of hematoxylin, Harris is the most commonly used due to the dark blue color providing a stark contrast against the pink eosin [19]. The protocol begins with deparaffinization and rehydration steps to remove the paraffin from the section and allow the stain to penetrate into the tissue. A representative image of an E15.5 tibia has been provided in Fig. 2.

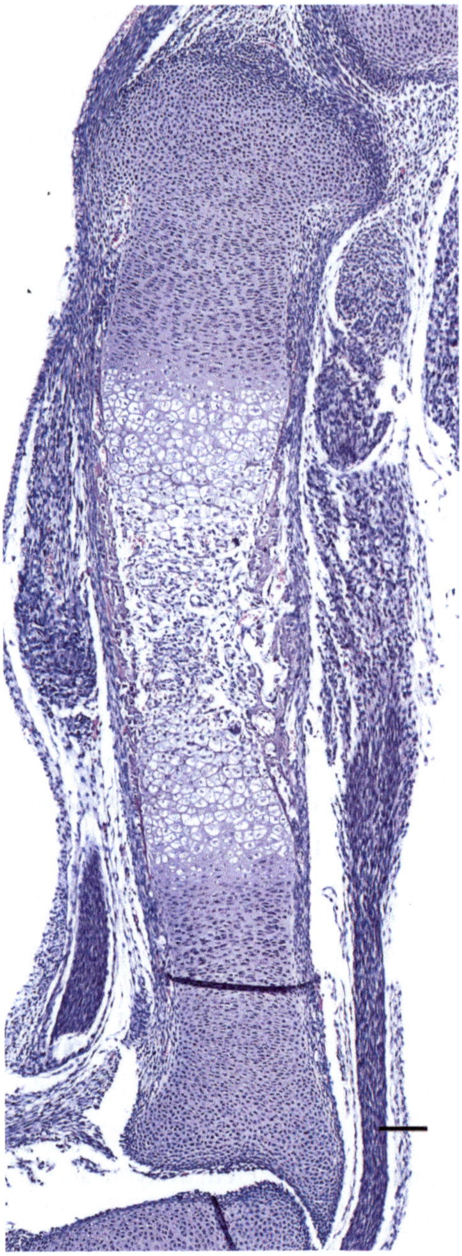

Fig. 2 Hematoxylin and eosin staining of an E15.5 murine tibia. Scale bar = 100 μm

1. Dry paraffin-embedded tissue sections on the slide warmer at 60 °C for at least 2–4 h (*see* **Note 17**).
2. Place the slides in a slide rack.
3. Incubate in xylene 3 times, 2 min each (*see* **Notes 18** and **20**).
4. Incubate in 100% ethanol 2 times, 1 min each.

5. Incubate in 95% ethanol 1 time for 1 min.
6. Incubate in distilled water 2 times, 1 min each.
7. Stain slides in Harris Hematoxylin for 75 s.
8. Gently rinse slides in cold tap water for 2 min (*see* **Note 19**).
9. Incubate in distilled water 2 times, 1 min each.
10. Incubate in 95% ethanol 1 time for 2 min.
11. Stain Slides in Eosin Y for 15 s.
12. Incubate in 95% ethanol 2 times, 30 s each.
13. Incubate in 100% ethanol 4 times, 1 min each.
14. Incubate in xylene 3 times, 2 min each.
15. Coverslip using xylene-based mounting media (*see* **Notes 21** and **22**).
16. Dry the slides overnight at room temperature on a slide tray.

3.4 Safranin O Staining for the Evaluation of GAGs in the Murine Growth Plate

Glycosaminoglycans (GAGs) are a critical component of the chondrocyte extracellular matrix. GAGs or mucopolysaccharides are homo- or heteropolymers consisting of a repeating disaccharide unit, which in turn consists of an amino sugar [*N*-acetylglucosamine or *N*-acetylgalactosamine and a uronic sugar (glucuronic acid or iduronic acid) or galactose [20]. GAGs are negatively charged, and therefore bind Safranin O, a red metachromatic cationic dye which binds the polyanions of the GAGs in a pH dependent fashion. This permits for the in vivo evaluation of the presence of GAGs in the growth plate [9, 14]. Figure 3 demonstrates an example of Safranin O staining on an E15.5 forelimb.

1. Dry paraffin-embedded tissue sections on the slide warmer at 60 °C for at least 2–4 h.
2. Place the slides in a slide rack.
3. Incubate in xylene 3 times, 2 min each (*see* **Notes 23** and **25**).
4. Incubate in 100% ethanol 2 times, 1 min each.
5. Incubate in 95% ethanol 1 time for 1 min.
6. Incubate in distilled water 1 time for 1 min.
7. Incubate in Weigert's iron hematoxylin working solution for 10 min.
8. Incubate in distilled water, 1 time for 1 min.
9. Rinse slides in running tap water for 10 min (*see* **Note 24**).
10. Incubate in distilled water, 1 time for 10–15 s.
11. Incubate slides in 0.1% Fast Green working solution for 10 min.
12. Rinse in 1% acetic water, no more than 10–15 s.

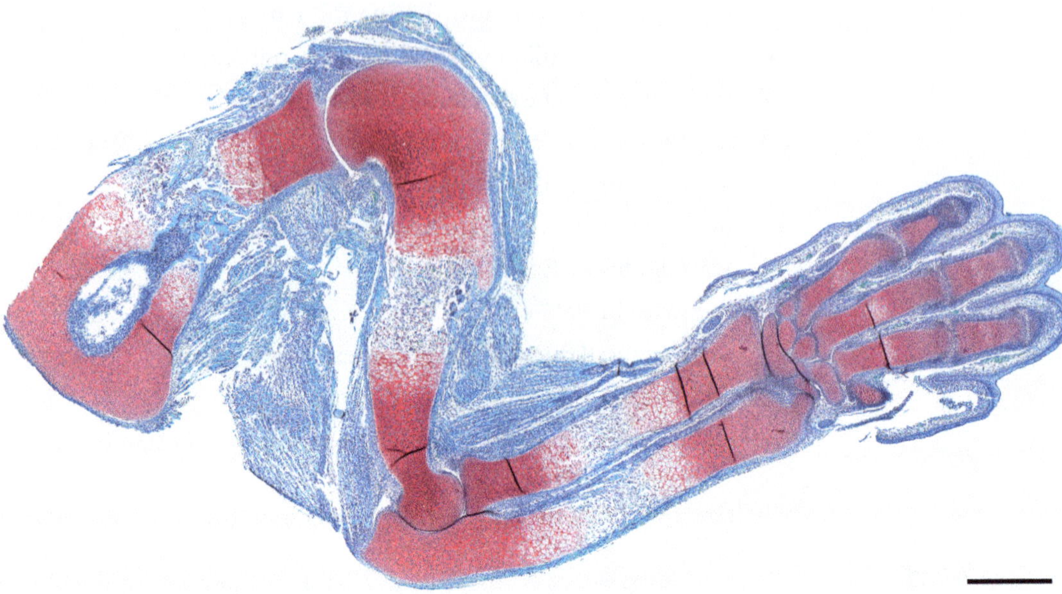

Fig. 3 Safranin O staining of an E15.5 murine forelimb. Note the GAG accumulation displayed in red. Scale bar = 500 μm

13. Incubate slides in 0.08% Safranin O working solution for 5 min.
14. Incubate in 95% ethanol 2 times, 30 s each.
15. Incubate in 100% ethanol 4 times, 1 min each.
16. Incubate in xylene 3 times, 2 min each.
17. Coverslip using xylene-based mounting media (*see* **Notes 26** and **27**).
18. Dry the slides overnight at room temperature on a slide tray.

3.5 In Situ Cell Death Detection (TUNEL Assay)

Terminal deoxynucleotidyl transferase dUTP nick-end labeling (TUNEL) is a method to detect DNA fragments and thus cell death. This staining utilizes the ability of the enzyme terminal deoxynucleotidyl transferase (TdT) to insert labeled dUTP, independently of a template, into free 3′-hydroxyl termini generated by the fragmentation of genomic DNA. When this ability is used to insert labeled nucleotides (dUTP), nuclei can be easily detected by standard immunohistochemical or immunofluorescent techniques [21, 22]. Immunohistochemical techniques can be both unwieldy and time consuming; we therefore prefer to use a direct immunofluorescent technique [7, 9, 14–17]. This protocol is derived from the Roche "In Situ Cell Death Detection Kit." Figure 4 displays a representative TUNEL assay on an E15.5 murine humerus proximal growth plate.

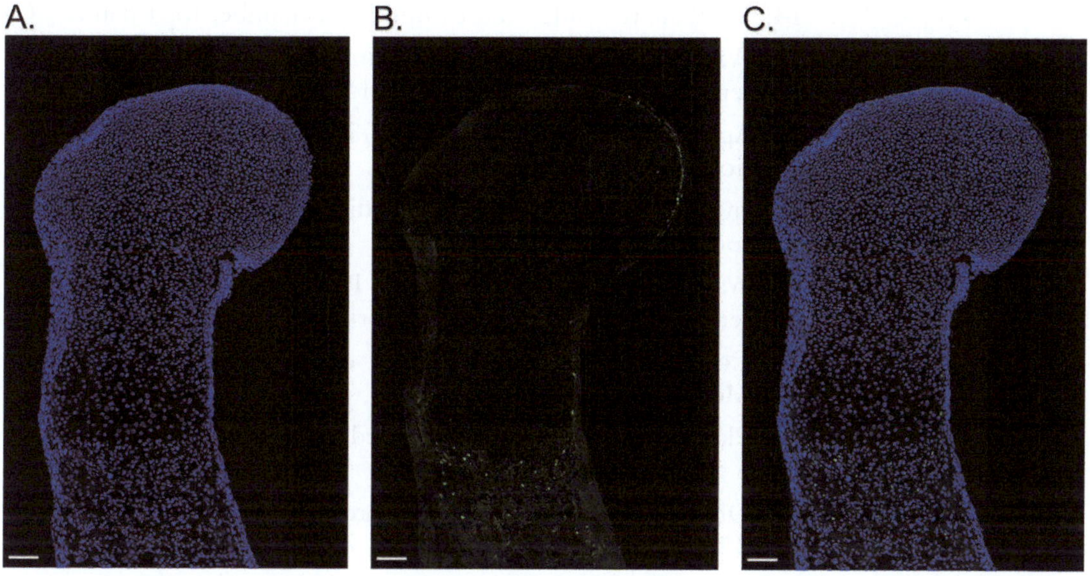

Fig. 4 TUNEL assay performed on an E15.5 murine humerus proximal growth plate. DAPI (**a**), TUNEL positive cells (**b**), and merged image (**c**). Note the presence of green TUNEL positive cells in the perichondrium and transition from hypertrophic layer to new bone. Scale bar = 100 μm

1. Dry paraffin-embedded tissue sections on the slide warmer at 60 °C overnight.
2. Prepare humidified chambers by covering the bottom with Whatman paper and adding 50 mL of 1× PBS onto the paper.
3. Prepare 250 mL of permeabilization solution.
4. Place the slides in a slide rack.
5. Incubate in xylene 3 times, 2 min each.
6. Incubate in 100% ethanol 2 times, 1 min each.
7. Incubate in 95% ethanol 1 time for 1 min.
8. Incubate in 1× PBS, 1 time for 5 min.
9. Incubate the slide in permeabilization solution by completely submerging them on ice for 2 min.
10. 10. Wash the slides 3 times, each with 250 mL of 1× PBS on a rocking platform for 2 min each at room temperature.
11. Dry the area around the tissue with a laboratory wipe and circle the tissue with a PAP pen (*see* **Note 28**).
12. Lay the slides flat in the prepared humidified chamber.
13. Apply approximately 50 μL of Roche In Situ Cell Death Detection Kit, Fluorescein Reaction Mixture to each slide. Additionally, apply the same volume of solution 2 only to the negative control slide. After this all slides need to be protected from light (*see* **Note 29**).

14. Incubate the slides in the humidified chamber for 1 h at 37 °C.
15. Wash the slides 3 times, for 3 min each, in 1× PBS at room temperature on a rocking platform.
16. Apply approximately 300 μL of 300 nM DAPI working solution to each slide (*see* **Note 30**).
17. Incubate the slides in the humidified chamber for 2–5 min at room temperature.
18. Wash the slides 3 times in 1× PBS for 2 min each at room temperature on a rocking platform.
19. Coverslip with 200–250 μL of Prolong Gold Antifade Mountant.
20. Place the slides in a foil wrapped slide box and store them at 4 °C (*see* **Note 31**).
21. Observe the sections with fluorescent microscope (*see* **Note 32**).

3.6 Immunohistochemistry on Formalin-Fixed, Paraffin-Embedded Tissues

Immunohistochemistry makes it possible to detect antigens in tissue sections using specific antigen-antibody interactions that ultimately result in the attachment of a marker to the antigen [23]. This marker may be a fluorescent dye (i.e., rhodamine, FITC, Hoechst) or the product of an enzymatic reaction (alkaline phosphatase, horseradish peroxidase: HRP) [24]. There is a broad array of histochemical techniques which can be used to localize antigens. The direct method is a single step technique that requires a labeled antibody to directly react with the antigen of interest [25, 26]. The indirect method, utilized in the following protocol, involves an unlabeled primary antibody binding to the antigen and a labeled secondary antibody detecting the primary [26, 27]. The secondary antibody can be labeled with an enzyme, fluorescent dye, or biotin. The use of biotin requires a further incubation with appropriately labeled streptavidin [28]. Formalin-fixed, paraffin-embedded tissues may be used for immunohistochemical techniques, but require a series of step to expose the epitopes for antibody binding. These include deparaffinization as well as antigen retrieval, which breaks the crosslinks created during formalin fixation using heat, pH, or proteases (i.e., hyaluronidase, chondroitinase, or proteinase K) [29]. To make use of an HRP-conjugated antibody it is necessary to prevent background staining caused by the presence of endogenous peroxidases, which can be quenched by treating the sections with a 3% solution of hydrogen peroxide/methanol. Furthermore, a blocking step is necessary to prevent nonspecific binding of the primary antibody. The most commonly utilized blocking agents include normal serum, nonfat dry milk, and bovine serum albumin (BSA). For some antibodies, signal amplification is a requirement. Numerous methodologies have been devised to amplify the signal, such as the Renaissance Tyramide Signal

Amplification (TSA) system (Perkin Elmer). This system uses an HRP-dependent catalysis and deposition of biotin-labeled tyramide onto tissue sections with a minimal loss of resolution [30]. The deposited biotin is then detected through the use of HRP-streptavidin. In this protocol, we illustrate an example of indirect immunohistochemistry using the TSA technology, useful in situations with particularly weak signals [14–17].

1. Dry paraffin-embedded tissue sections on the slide warmer at 60 °C overnight.
2. Prepare humidified chambers by covering the bottom of the chamber with Whatman paper and distributing 50 mL of 1× TBS onto the paper.
3. Place the slides in a slide rack.
4. Incubate in xylene 3 times, 5 min each.
5. Incubate in 100% ethanol 2 times, 1 min each.
6. Incubate in 95% ethanol 1 time for 1 min.
7. Incubate in 1× TBS, 1 time for 5 min.
8. Incubate the slides in a container of 250 mL of sodium citrate solution for 10 min in a 95 °C water bath (*see* **Note 34**).
9. Remove the container from the water bath and allow to sit at room temperature for 30 min.
10. Incubate the slides in 250 mL of 3% H_2O_2/MeOH for 10 min at room temperature.
11. Wash the slides in 250 mL distilled water for 5 min at room temperature.
12. Wash the slides in 250 mL of 1× TBS for 1 min at room temperature.
13. Using a laboratory wipe, dry the area around the tissue on the slides and circle tissue sections with a PAP pen.
14. Lay the slides flat in the humidified chamber.
15. Apply approximately 100 μL of TNB onto each slide.
16. Incubate the slides in the humidified chamber for 30 min at room temperature.
17. Pour off TNB and immediately add 100 μL of unconjugated primary antibody (diluted to manufacturer's recommended concentration in TNB). Apply only TNB to the negative control sections (*see* **Notes 33** and **35**).
18. Incubate the slides in the humidified chamber overnight at 4 °C.
19. On the following day, allow the slides to equilibrate to room temperature for 30 min.

20. Wash the slides, 3 times for 2 min each, in 1× TNT on a rocking platform at room temperature.
21. Apply 100 μL of the appropriate biotinylated secondary antibody (diluted to manufacturer's recommended concentration in TNB).
22. Incubate the slides in the humidified chamber for 30 min at room temperature.
23. Wash the slides, 3 times for 2 min each, in 1× TNT on a rocking platform at room temperature.
24. Apply 100 μL HRP-Streptavidin (HRP-SA from TSA kit) diluted 1:1000 in TNB.
25. Incubate the slides for 30 min at room temperature in the humidified chamber.
26. Wash the slides, 3 times for 2 min each, in 1× TNT on a rocking platform at room temperature.
27. Apply 100 μL biotinyl tyramide (from TSA kit) diluted 1:50 in the amplification diluent provided with the kit.
28. Incubate the slides in the humidified chamber at room temperature for 5 min.
29. Wash the slides, 3 times for 2 min each, in 1× TNT on a rocking platform at room temperature.
30. Apply 100 μL DAB (3,3′-diaminobenzidine) peroxidase (HRP) substrate kit Solution on each slide (*see* **Notes 36** and **37**).
31. Incubate the slides in the humidified chamber until brown color develops.
32. Rinse the slides in distilled water.
33. Counterstain the slides with two drops of Mayer's hematoxylin.
34. Incubate the slides in the humidified chamber for 3 min at room temperature.
35. Wash the slides under cold running tap water for 2 min.
36. Incubate the slides in 250 mL 1× PBS for 30 s.
37. Wash the slides in 250 mL distilled water for 1 min.
38. Incubate the slides in 95% ethanol, 2 times for 1 min each.
39. Incubate the slides in 100% ethanol, 4 times for 1 min each.
40. Incubate the slides in xylene, 3 times for 2 min each.
41. Coverslip the slides with xylene-based mounting media.
42. Dry the slides overnight on a slide tray at room temperature.

3.7 EdU Assay for Evaluating Cellular Proliferation

A major disadvantage of traditional proliferation techniques, such as BrdU, is that it requires the use of trypsin (or similar reagents), which can result in damage to the tissue. Furthermore, BrdU staining is a time-intensive procedure. A simpler and more rapid proliferation assay, EdU, has been developed, which avoids these pitfalls [31]. It makes use of 5-ethynyl-2′-deoxyuridine (EdU), a nucleoside analog of thymidine which can be incorporated into the DNA during S-phase. Detection of the included EdU is based on a copper-catalyzed cycloaddition, where a stable triazole ring is formed by covalently coupling the alkaline group present in the EdU and the Alexa Fluor–conjugated azide group. This technique requires the injection of EdU intraperitoneally into a pregnant mouse. An example of an EdU assay on an E15.5 murine humerus proximal growth plate is shown in Fig. 5.

1. Weigh the pregnant mouse.
2. Using a 1 mL syringe fitted a 27-G needle, inject the EdU stock solution intraperitoneally 2 h before sacrificing the mouse. Inject 0.1 mL of 10 mM EdU stock solution per 10 g of mouse (*see* **Note 38**).
3. Deliver the embryos through cesarean section as described in the embryo harvest protocol (Subheading 3.1).
4. Place the embryos in a 100 mm petri dish of 1× PBS on ice.
5. Separate the embryos and thoroughly rinse them with sterile 1× PBS.

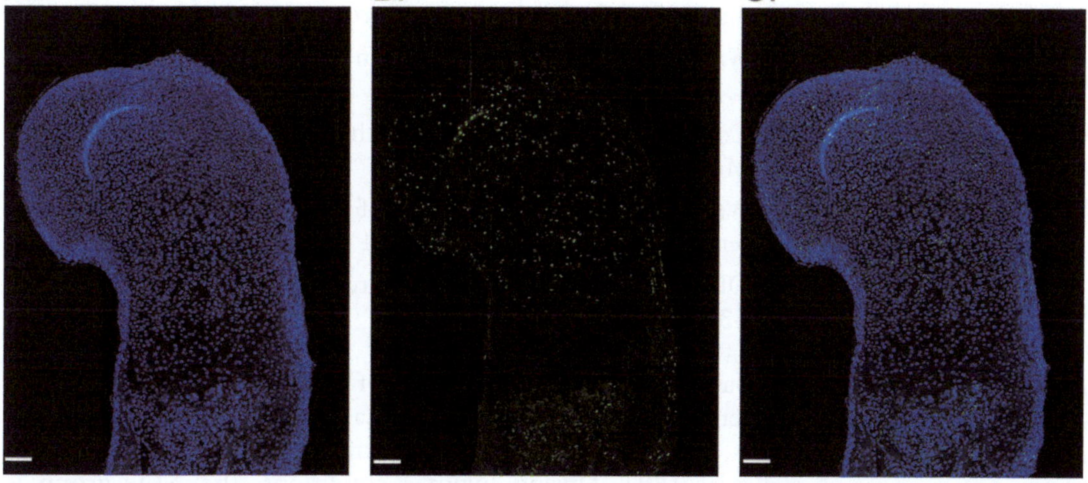

Fig. 5 EdU assay of an E15.5 murine humerus proximal growth plate. DAPI (**a**), EdU positive cells (**b**), and merged image (**c**). Note the presence of green EdU positive cells throughout the round and columnar proliferative layers of chondrocytes. Scale bar = 100 μm

6. Rinse each embryo 3 times with 10 mL sterile 1× PBS in a clean 100 mm petri dish, cleaning instruments with 70% ethanol between different embryos.
7. Fix, dissect, process, and section specimens as described in the initial protocol (Subheading 3.1).
8. Dry paraffin-embedded tissue sections on the slide warmer at 60 °C overnight.
9. Prepare humidified chambers by covering the bottom with Whatman paper and adding 50 mL of 1× PBS onto the paper.
10. Place the slides in a slide rack.
11. Incubate in xylene 3 times, 2 min each.
12. Incubate in 100% ethanol 2 times, 1 min each.
13. Incubate in 95% ethanol 1 time for 1 min.
14. Incubate in 1× PBS, 1 time for 5 min.
15. Using a laboratory wipe, dry the area around each tissue section without overdrying the tissue itself.
16. Circle the tissue section with a PAP pen.
17. Lay the slides flat in the humidified chamber.
18. Apply 100–200 μL of the Click-iT EdU Alexa Fluor 488 Imaging Kit Reaction Mixture to each section and incubate in the humidified chamber for 30 min at room temperature protected from light (*see* **Notes 39** and **40**).
19. Wash the slides 2 times for 2 min each at room temperature in 1× PBS.
20. Apply 100–200 μL Hoechst 33342 working solution to each section and incubate for 2 min at room temperature in the humidified chamber protected from light (*see* **Note 41**).
21. Wash the slides 2 times for 2 min each at room temperature in 1× PBS.
22. Coverslip the slides using the Prolong Gold Antifade Mountant.
23. Wrap the slides in aluminum foil and store them at room temperature for 24 h and then 4 °C (*see* **Note 42**).
24. Observe the sections using fluorescent microscopy (*see* **Notes 43** and **44**).

3.8 EF5 Staining to Detect Hypoxia in the Developing Growth Plate

EF5, a pentafluorinated derivative of etanidazole developed at the University of Pennsylvania by Dr. Cameron Koch and Dr. Sydney Evans, has a nitro (NO_2) group attached to the imidazole ring structure [32]. During hypoxic conditions, the NO_2 group is reduced to an amino group (NH_2) [33, 34]. One of the highly reactive intermediates of this reduction has the ability to bind to any cellular protein, forming adducts which can then be recognized

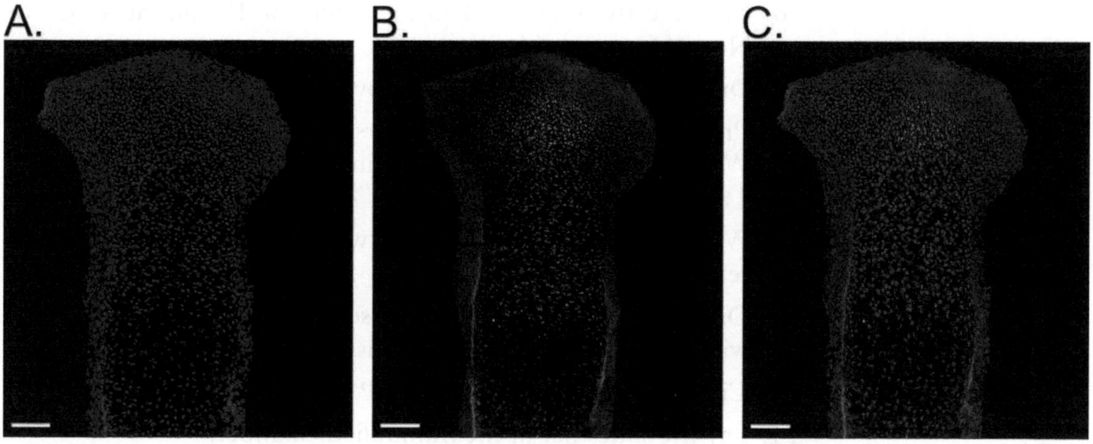

Fig. 6 EF5 assay performed on an E15.5 murine tibial proximal growth plate. DAPI (**a**), EF5 (**b**), and merged image (**c**). Note the red fluorescence which denotes the central hypoxic region within the growth plate. Scale bar = 100 μm

through the use of specific antibodies [32–34]. Due to this, EF5 is a marker which can be useful in detecting hypoxia both in vivo and in vitro [33, 34]. As the chondrocytes of the developing growth plate both survive and differentiate in a hypoxic environment, changes in the oxygen levels of cartilage are critical for its development [9, 16, 17, 35–37]. We analyze levels of hypoxia in the growth plate through the use of an immunofluorescent detection of EF5 adducts, which requires the intraperitoneal injection of EF5 into a pregnant mouse [16, 17]. A representative picture of EF5 staining on an E15.5 murine tibia proximal growth plate is displayed in Fig. 6.

1. Weigh the pregnant mouse.
2. Using the 1 mL syringe with a 27 G needle, inject the 10 mM EF5 stock solution 2 h prior to sacrifice. Injection should be 0.1 mL of 10 mM EF5 stock solution per 10 g of mouse intraperitoneally (*see* **Note 45**).
3. Deliver the embryos via cesarean section, as described previously (Subheading 3.2).
4. Place the embryos in a 100 mm petri dish on ice.
5. Dissect the embryos and thoroughly rinse with sterile 1× PBS in a clean 100 mm petri dish.
6. Rinse each embryo 3 times individually with sterile 1× PBS in a succession of clean petri dishes. Clean instruments with 70% ethanol between each embryo.
7. Dissect, embed, and section specimens as described in the fresh frozen specimen isolation protocol (Subheading 3.2).

8. Incubate the slides in 100% acetone for 10 min at 4 °C (*see* **Note 46**).
9. Dry the freshly frozen sections overnight at room temperature.
10. Prepare the humidified chambers by covering the bottom with Whatman paper and distributing approximately 50 mL of 1× PBS.
11. Wash the sections in 1× PBS twice, each for 2 min at room temperature (*see* **Note 47**).
12. Dry the area around the tissue section with a laboratory wipe without overdrying the actual tissue.
13. Circle tissue sections with a PAP pen.
14. Lay the slides flat in the humidified chamber.
15. Apply 100–200 μL of 5% mouse serum–PBS blocking solution to each section.
16. Incubate the slides in the humidified chamber for 30 min at room temperature.
17. Prepare the anti-EF5 conjugated antibody by diluting 1:25 in 3% bovine serum albumin–PBS while protecting from light.
18. Apply 100–200 μL of the solution onto each section (*see* **Notes 48–50**).
19. Incubate the slides in the humidified chamber for 2 h at room temperature while protecting from light.
20. Wash the sections twice for 2 min each in 1× PBS at room temperature while protected from light.
21. Rinse the slides in distilled water for 30 s.
22. Apply 100–200 μL 300 nM DAPI working solution to each section (*see* **Note 51**).
23. Incubate the slides in the humidified chamber for 5 min at room temperature while protecting from light.
24. Coverslip the slides using 200–250 μL Prolong Gold Antifade Mountant.
25. Wrap the slides in aluminum foil and store them at room temperature for 24 h and then move to 4 °C (*see* **Note 52**).
26. Observe the sections with fluorescent microscopy (*see* **Note 53**).

3.9 Fluorescent Image Quantification Using ImageJ

Quantitative assessment of the specific information yielded by staining and labeling techniques is crucial to fully understand growth plate development. Automated techniques and image analysis tools have been developed to perform this quantification. In general, the principle behind these tools consists of first designating a region of interest within which the total number of cells can be calculated.

Following this, cells exhibiting a positive signal within the same region of interest will also be counted. This allows for the precise determination of the ratio of positive cells to the total cell number.

1. During image acquisition, be sure to save two images of the same field. The first using the DAPI channel to allow for the quantification of total cell number. The second using the channel corresponding to your fluorophore of interest.
2. Open both files in Adobe Photoshop.
3. Create a new file that is the size of your images (File/new...).
4. Create three distinct layers in the new file (Layer/new/Layer...).
5. Copy and paste the DAPI image onto the first layer.
6. Copy and paste the fluorescent image onto the second layer.
7. On the third layer, using the rectangle or ellipse tool, outline your region of interest. Save the ROI overlaid on the fluorescent signal image as a .jpeg file (*see* **Note 54**).
8. Change the order of the layers so the region of interest is shown superimposed on the DAPI image, and save this as a .jpeg as well.
9. Open ImageJ.
10. Open the DAPI + region of interest file.
11. Select the region of interest using the rectangle or ellipse tool, and copy it to a new file.
12. Go to Image, Adjust, Brightness/Contrast, Auto, and save.
13. Go to Process, Binary, Make Binary (this will produce a binary black and white image).
14. Go to Process, Binary, Watershed (this divides two adjacent cells that were originally designated as one).
15. Go to analyze, analyze particles, and select the following options:
    (a) Size (pixel^2): 5—Infinity, so that noise is eliminated.
    (b) Circularity: 0.00–1.00, so that different shapes from completely elongated to perfectly circular are included.
    (c) Show ellipses, so that you can see where and which cells are identified by the program.
    (d) Check the boxes "Display results" (displays individual results as a table that can be saved as an Excel file), "Summarize" (summarized results will also be saved as a table that can be made into an Excel file), and "Add to manager" (plots the cells that the program identifies).
16. Apply the same sequence of steps to the fluorescent signal image.

4 Notes

1. The cross-linking resulting from paraformaldehyde fixation provides excellent accessibility and retention of the RNA in the tissue.
2. For optimal results, dissection and processing should be performed within 1–2 weeks following fixation.
3. All processing steps are performed in an automated processor. The volumes and ethanol and xylene used with depend on the type and model of automated processor used.
4. The ethanol gradient during processing progressively dehydrates the tissue. This is important because paraffin is a hydrophobic wax, and the presence of water in the specimen will prevent a full infiltration of the embedding media. It is conducted in a stepwise fashion to prevent tissue distortion.
5. Xylene is miscible with both ethanol and paraffin, so it is an ideal intermediate solvent for clearing ethanol from the tissues and preparing them for paraffin infiltration.
6. The correct temperature for the paraffin embedding steps depends on the type of paraffin used. We use a "low" melting temperature paraffin (below 60 °C) as this generally better preserves epitopes and mRNA integrity in the tissue.
7. Proper orientation of specimens in the mold while embedding is crucial for obtaining quality sections. Our orientation methods for embedding were devised to overcome the natural curvature found in these bony parts, preventing misinterpretation during growth plate analysis. We orient specimens as follows: forelimbs with the dorsal side down, pads facing up, hind limbs with the inner thigh facing down, and spines with the dorsal side facing down. Right forelimbs should be oriented so that the scapula or humerus is pointing to the right and the distal ends of the radius and ulna point toward the left of the mold. We embed left forelimbs in the opposite orientation. Right hind limbs should be embedded so the femur points to the right and the paw to the left of the mold. The opposite is true of left hind limbs.
8. While embedding, feel the back of the block before removing from the mold. It should feel cold to the touch. If it is not yet cold, leave the block on the cold plate to chill further. This will make it easier to remove the block from the mold, and reduce the risk of damage as it is removed.
9. While sectioning, frequently cool the block in the ice bath for the best quality sections.

10. While orienting the block in the microtome chuck, have the proximal and distal growth plates at the same level (parallel section).
11. Always place the block back into the chuck with the same orientation. This can be kept track of by removing a small portion of one corner of the block with a razor, and always placing the cut corner back in the same position in the chuck. This also helps in separating floating sections in the water bath.
12. Isopentane chilled with dry ice can also be used to freeze the OCT blocks.
13. Immediately store the frozen blocks at −80 °C. Frozen blocks can be stored in this manner for up to 1 year.
14. Orient the blocks in the plastic mold in an identical fashion to what was previously described for paraffin embedding.
15. Orient the block in the cryostat chuck so the proximal and distal growth plates are at the same level.
16. Frozen sections can be stored for up to 2 weeks at −20 °C and up to 6 months at −80 °C.
17. H&E staining can be performed on frozen sections by avoiding the Deparaffinization and dehydration steps. Wash 2 times, 1 min each in 1× phosphate buffered saline and 1 time in distilled water for 30 s. Then proceed directly to the Harris Hematoxylin step.
18. Xylene, 100% ethanol, 95% ethanol, Harris Hematoxylin, and Eosin Y can be reused for 3 to 4 times. Harris Hematoxylin must be filtered before each us.
19. Let the tap water run for a few minutes before use to allow it to cool down. Ensure the water does not run directly onto slides.
20. Ensure Coplin jars are sufficiently full to completely immerse the tissue sections.
21. If bubbles form during coverslipping, gently press on the glass coverslip with forceps to force them out.
22. Before applying the coverslip, cleaning the back of the slide with a Kimwipes can remove excess stain and reduce background while imaging.
23. Ensure all reagents are new and clean before every staining. Do not reuse.
24. Let the tap water run for a few minutes before use to allow it to cool down. Ensure the water does not run directly onto slides.
25. Ensure Coplin jars are sufficiently full to completely immerse the tissue sections.
26. If bubbles form during coverslipping, gently press on the glass coverslip with forceps to force them out.

27. Before applying the coverslip, cleaning the back of the slide with a Kimwipes can remove excess stain and reduce background while imaging.
28. Use a laboratory wipe to dry the area around the tissue before circling with the PAP pen, but avoid overdrying the actual tissue.
29. A negative control should be included with each experiment to test for nonspecific binding of the labeled dUTPs.
30. DAPI is a potential mutagen and teratogen. Appropriate protective equipment should be worn, and the solutions discarded in a proper manner in keeping with their biohazardous nature.
31. Stained slides can be stored up to 2 weeks at 4 °C or 6 months at −20 °C if well protected from light. However, for best results, retrieve the fluorescence within 7 days of staining.
32. Use an excitation wavelength of 350–400 nm and emission wavelength of 450–500 nm (blue) to detect DAPI stained nuclei. TUNEL-positive cells can be visualized using an excitation wavelength of 450–500 nm range and an emission wavelength of 515–565 nm (green).
33. Include negative control sections that do not receive primary antibody in every experiment.
34. Do not begin the antigen retrieval step until the sodium citrate solution has reached 95 °C.
35. During antibody steps, keep all reagents on ice at all times.
36. Protect DAB solution from light and use within 1 h.
37. If necessary, two drops of nickel solution from the DAB HRP substrate kit can be added to the DAB solution to obtain a darker signal.
38. EdU is a potential mutagen and teratogen. It should be handled and disposed of in accordance with its biohazardous nature and institutional regulations.
39. Prepare the Click-iT reaction mixture in the specific order described above and use within 15 min of preparation for best results. This volume (500 μL) can stain ten small tissue sections.
40. After adding the Click-iT reaction mixture, perform all following steps in the dark protecting the sections from light.
41. 300 nM DAPI working solution can be used instead of the Hoechst 33342 working solution. Prepare the DAPI solution by beginning with a 5 mg/mL stock solution by dissolving 10 mg of DAPI hydrochloride in 2 mL of distilled water. This can be divided into 50 μL aliquots and stored up to 2 years at −20 °C. Prepare the 300 nM working solution by diluting the stock solution 1:5000 in 1× phosphate buffered saline. This

can be stored up to 6 weeks at 4 °C protected from light. Both DAPI and Hoechst are potential mutagens and teratogens, and should be handled and disposed of accordingly.

42. Stained slides can be stored up to 2 weeks at 4 °C or up to 6 months at −20 °C if protected from light.
43. Use an excitation wavelength between 350 and 400 nm and an emission wavelength of 450–500 nm (blue) to detect Hoechst 33342 stained nuclei (or DAPI) and an excitation wavelength between 450 and 500 nm and an emission wavelength in the range of 515–565 nm (green) to detect the Alexa Fluor 488 EdU positive cells.
44. Retrieve the fluorescent signal within 7 days for best results.
45. The EF5 powder and antibodies required must be directly purchased from Dr. Cameron Koch (University of Pennsylvania Department of Radiation Oncology) at the following site: http://hypoxia-imaging.org.
46. Place the acetone at 4 °C a few hours before the staining to ensure it is sufficiently cooled.
47. Do not perform any of the wash steps on a rocking platform and be gentle transferring the sections from solution to solution as frozen sections are fragile and tend to detach from the slide easily.
48. Include a negative control for all staining experiments. For this staining, the negative control should receive only the 3% BSA–PBS without the anti-EF5 conjugated antibody added.
49. Protect the slides from light after incubating with the anti-EF5 conjugated antibody.
50. The anti-EF5 antibody can be conjugated with the following substrates: biotin, Cy3, Cy5, or Alexa 488. We recommend the use of an antibody conjugated with a fluorescent dye for best results.
51. DAPI is a potential mutagen and teratogen, and should be handled in accordance with its biohazardous nature and institutional regulations.
52. Stained slides can be stored up to 2 weeks at 4 °C or up to 6 months at −20 °C if well protected from light. However, for best results, retrieve the fluorescence within 7 days.
53. Use an excitation wavelength of 350–400 nm and an emission wavelength of 450–500 nm (blue) to visualize the DAPI stained nuclei, and the correct excitation and emission wavelength for your selected dye conjugated to the anti-EF5 antibody.

54. It is of the utmost importance to consistently and appropriately define the region of interest. This is because the proportion of cells exhibiting a positive signal or the intensity of that signal may vary from one tissue to another as well as within different regions of the same tissue.

References

1. Karsenty G (2003) The complexities of skeletal biology. Nature 423(6937):316–318. https://doi.org/10.1038/nature01654
2. Kronenberg HM (2003) Developmental regulation of the growth plate. Nature 423 (6937):332–336. https://doi.org/10.1038/nature01657
3. Michigami T (2013) Regulatory mechanisms for the development of growth plate cartilage. Cell Mol Life Sci 70(22):4213–4221. https://doi.org/10.1007/s00018-013-1346-9
4. Zelzer E, Olsen BR (2003) The genetic basis for skeletal diseases. Nature 423 (6937):343–348. https://doi.org/10.1038/nature01659
5. Aigner T, Stove J (2003) Collagens—major component of the physiological cartilage matrix, major target of cartilage degeneration, major tool in cartilage repair. Adv Drug Deliv Rev 55(12):1569–1593
6. Heinegard D (2009) Fell-Muir Lecture: proteoglycans and more—from molecules to biology. Int J Exp Pathol 90(6):575–586. https://doi.org/10.1111/j.1365-2613.2009.00695.x
7. Maes C, Carmeliet G, Schipani E (2012) Hypoxia-driven pathways in bone development, regeneration and disease. Nat Rev Rheumatol 8(6):358–366. https://doi.org/10.1038/nrrheum.2012.36
8. Zelzer E, Mamluk R, Ferrara N, Johnson RS, Schipani E, Olsen BR (2004) VEGFA is necessary for chondrocyte survival during bone development. Development 131 (9):2161–2171. https://doi.org/10.1242/dev.01053
9. Yao Q, Khan MP, Merceron C, LaGory EL, Tata Z, Mangiavini L, Hu J, Vemulapalli K, Chandel NS, Giaccia AJ, Schipani E (2019) Suppressing mitochondrial respiration is critical for hypoxia tolerance in the fetal growth plate. Dev Cell 49(5):748–763.e747. https://doi.org/10.1016/j.devcel.2019.04.029
10. Karsenty G, Ferron M (2012) The contribution of bone to whole-organism physiology. Nature 481(7381):314–320. https://doi.org/10.1038/nature10763
11. Mouse Genome Sequencing C, Waterston RH, Lindblad-Toh K, Birney E, Rogers J, Abril JF, Agarwal P, Agarwala R, Ainscough R, Alexandersson M, An P, Antonarakis SE, Attwood J, Baertsch R, Bailey J, Barlow K, Beck S, Berry E, Birren B, Bloom T, Bork P, Botcherby M, Bray N, Brent MR, Brown DG, Brown SD, Bult C, Burton J, Butler J, Campbell RD, Carninci P, Cawley S, Chiaromonte F, Chinwalla AT, Church DM, Clamp M, Clee C, Collins FS, Cook LL, Copley RR, Coulson A, Couronne O, Cuff J, Curwen V, Cutts T, Daly M, David R, Davies J, Delehaunty KD, Deri J, Dermitzakis ET, Dewey C, Dickens NJ, Diekhans M, Dodge S, Dubchak I, Dunn DM, Eddy SR, Elnitski L, Emes RD, Eswara P, Eyras E, Felsenfeld A, Fewell GA, Flicek P, Foley K, Frankel WN, Fulton LA, Fulton RS, Furey TS, Gage D, Gibbs RA, Glusman G, Gnerre S, Goldman N, Goodstadt L, Grafham D, Graves TA, Green ED, Gregory S, Guigo R, Guyer M, Hardison RC, Haussler D, Hayashizaki Y, Hillier LW, Hinrichs A, Hlavina W, Holzer T, Hsu F, Hua A, Hubbard T, Hunt A, Jackson I, Jaffe DB, Johnson LS, Jones M, Jones TA, Joy A, Kamal M, Karlsson EK, Karolchik D, Kasprzyk A, Kawai J, Keibler E, Kells C, Kent WJ, Kirby A, Kolbe DL, Korf I, Kucherlapati RS, Kulbokas EJ, Kulp D, Landers T, Leger JP, Leonard S, Letunic I, Levine R, Li J, Li M, Lloyd C, Lucas S, Ma B, Maglott DR, Mardis ER, Matthews L, Mauceli E, Mayer JH, McCarthy M, McCombie WR, McLaren S, McLay K, McPherson JD, Meldrim J, Meredith B, Mesirov JP, Miller W, Miner TL, Mongin E, Montgomery KT, Morgan M, Mott R, Mullikin JC, Muzny DM, Nash WE, Nelson JO, Nhan MN, Nicol R, Ning Z, Nusbaum C, O'Connor MJ, Okazaki Y, Oliver K, Overton-Larty E, Pachter L, Parra G, Pepin KH, Peterson J, Pevzner P, Plumb R, Pohl CS, Poliakov A, Ponce TC, Ponting CP, Potter S, Quail M, Reymond A, Roe BA, Roskin KM, Rubin EM, Rust AG, Santos R, Sapojnikov V, Schultz B, Schultz J, Schwartz MS, Schwartz S, Scott C, Seaman S, Searle S, Sharpe T, Sheridan A, Shownkeen R, Sims S, Singer JB, Slater G, Smit A, Smith DR,

Spencer B, Stabenau A, Stange-Thomann N, Sugnet C, Suyama M, Tesler G, Thompson J, Torrents D, Trevaskis E, Tromp J, Ucla C, Ureta-Vidal A, Vinson JP, Von Niederhausern AC, Wade CM, Wall M, Weber RJ, Weiss RB, Wendl MC, West AP, Wetterstrand K, Wheeler R, Whelan S, Wierzbowski J, Willey D, Williams S, Wilson RK, Winter E, Worley KC, Wyman D, Yang S, Yang SP, Zdobnov EM, Zody MC, Lander ES (2002) Initial sequencing and comparative analysis of the mouse genome. Nature 420(6915):520–562. https://doi.org/10.1038/nature01262
12. Nguyen D, Xu T (2008) The expanding role of mouse genetics for understanding human biology and disease. Dis Model Mech 1(1):56–66. https://doi.org/10.1242/dmm.000232
13. Piret SE, Thakker RV (2011) Mouse models for inherited endocrine and metabolic disorders. J Endocrinol 211(3):211–230. https://doi.org/10.1530/JOE-11-0193
14. Mangiavini L, Merceron C, Araldi E, Khatri R, Gerard-O'Riley R, Wilson TL, Rankin EB, Giaccia AJ, Schipani E (2014) Loss of VHL in mesenchymal progenitors of the limb bud alters multiple steps of endochondral bone development. Dev Biol 393(1):124–136. https://doi.org/10.1016/j.ydbio.2014.06.013
15. Pfander D, Kobayashi T, Knight MC, Zelzer E, Chan DA, Olsen BR, Giaccia AJ, Johnson RS, Haase VH, Schipani E (2004) Deletion of Vhlh in chondrocytes reduces cell proliferation and increases matrix deposition during growth plate development. Development 131(10):2497–2508. https://doi.org/10.1242/dev.01138
16. Provot S, Schipani E (2007) Fetal growth plate: a developmental model of cellular adaptation to hypoxia. Ann N Y Acad Sci 1117:26–39. https://doi.org/10.1196/annals.1402.076
17. Schipani E, Ryan HE, Didrickson S, Kobayashi T, Knight M, Johnson RS (2001) Hypoxia in cartilage: HIF-1alpha is essential for chondrocyte growth arrest and survival. Genes Dev 15(21):2865–2876. https://doi.org/10.1101/gad.934301
18. Titford M (2005) The long history of hematoxylin. Biotech Histochem 80(2):73–78. https://doi.org/10.1080/10520290500138372
19. Levdik TI (1989) Unification of the staining of histological preparations and histoautoradiograms with Harris hematoxylin. Arkh Patol 51(7):81–82
20. Lindahl U, Couchman J, Kimata K, Esko JD (2015) Proteoglycans and sulfated glycosaminoglycans. In: Varki A, Cummings RD et al (eds) Essentials of glycobiology. Cold Spring Harbor Laboratory Press, Cold Spring Harbor (NY), pp 207–221. https://doi.org/10.1101/glycobiology.3e.017
21. Gavrieli Y, Sherman Y, Ben-Sasson SA (1992) Identification of programmed cell death in situ via specific labeling of nuclear DNA fragmentation. J Cell Biol 119(3):493–501. https://doi.org/10.1083/jcb.119.3.493
22. Loo DT, Rillema JR (1998) Measurement of cell death. Methods Cell Biol 57:251–264. https://doi.org/10.1016/s0091-679x(08)61583-6
23. Ramos-Vara JA, Miller MA (2014) When tissue antigens and antibodies get along: revisiting the technical aspects of immunohistochemistry—the red, brown, and blue technique. Vet Pathol 51(1):42–87. https://doi.org/10.1177/0300985813505879
24. Taylor CR, Shi S-R, Barr N, Wu N (2013) Techniques of immunohistochemistry: principles, pitfalls, and standardization. Diagn Immunohistochem 2:1–42
25. Coons AH, Kaplan MH (1950) Localization of antigen in tissue cells; improvements in a method for the detection of antigen by means of fluorescent antibody. J Exp Med 91(1):1–13. https://doi.org/10.1084/jem.91.1.1
26. Polak JM, Van Noorden S (2003) Introduction to immunocytochemistry, 3rd edn. BIOS Scientific Publishers, Oxford
27. Coons AH, Leduc EH, Connolly JM (1955) Studies on antibody production. I. A method for the histochemical demonstration of specific antibody and its application to a study of the hyperimmune rabbit. J Exp Med 102(1):49–60. https://doi.org/10.1084/jem.102.1.49
28. Guesdon JL, Ternynck T, Avrameas S (1979) The use of avidin-biotin interaction in immunoenzymatic techniques. J Histochem Cytochem 27(8):1131–1139. https://doi.org/10.1177/27.8.90074
29. Bogen SA, Vani K, Sompuram SR (2009) Molecular mechanisms of antigen retrieval: antigen retrieval reverses steric interference caused by formalin-induced cross-links. Biotech Histochem 84(5):207–215. https://doi.org/10.3109/10520290903039078
30. Gross AJ, Sizer IW (1959) The oxidation of tyramine, tyrosine, and related compounds by peroxidase. J Biol Chem 234(6):1611–1614
31. Salic A, Mitchison TJ (2008) A chemical method for fast and sensitive detection of DNA synthesis in vivo. Proc Natl Acad Sci U

S A 105(7):2415–2420. https://doi.org/10.1073/pnas.0712168105

32. Koch CJ (2002) Measurement of absolute oxygen levels in cells and tissues using oxygen sensors and 2-nitroimidazole EF5. Methods Enzymol 352:3–31. https://doi.org/10.1016/s0076-6879(02)52003-6
33. Horsman MR, Mortensen LS, Petersen JB, Busk M, Overgaard J (2012) Imaging hypoxia to improve radiotherapy outcome. Nat Rev Clin Oncol 9(12):674–687. https://doi.org/10.1038/nrclinonc.2012.171
34. Kizaka-Kondoh S, Konse-Nagasawa H (2009) Significance of nitroimidazole compounds and hypoxia-inducible factor-1 for imaging tumor hypoxia. Cancer Sci 100(8):1366–1373. https://doi.org/10.1111/j.1349-7006.2009.01195.x
35. Amarilio R, Viukov SV, Sharir A, Eshkar-Oren I, Johnson RS, Zelzer E (2007) HIF1alpha regulation of Sox9 is necessary to maintain differentiation of hypoxic prechondrogenic cells during early skeletogenesis. Development 134(21):3917–3928. https://doi.org/10.1242/dev.008441
36. Maes C, Araldi E, Haigh K, Khatri R, Van Looveren R, Giaccia AJ, Haigh JJ, Carmeliet G, Schipani E (2012) VEGF-independent cell-autonomous functions of HIF-1alpha regulating oxygen consumption in fetal cartilage are critical for chondrocyte survival. J Bone Miner Res 27(3):596–609. https://doi.org/10.1002/jbmr.1487
37. Provot S, Zinyk D, Gunes Y, Kathri R, Le Q, Kronenberg HM, Johnson RS, Longaker MT, Giaccia AJ, Schipani E (2007) Hif-1alpha regulates differentiation of limb bud mesenchyme and joint development. J Cell Biol 177(3):451–464. https://doi.org/10.1083/jcb.200612023

Chapter 6

Preparation of Adult Mouse Skeletal Tissue Sections for RNA In Situ Hybridization

Charles de Charleroy, Abdul Haseeb, and Véronique Lefebvre

Abstract

The RNA in situ hybridization assay is essential in many studies to evaluate gene expression in vivo. It consists of generating tissue sections and subsequently hybridizing these sections with RNA probes. Keeping RNA intact is a challenge while harvesting tissue samples, processing through embedding, sectioning them, and conditioning the sections for hybridization. These challenges are particularly strong for adult skeletal tissues due to their copious, dense, and mineralized extracellular matrices. Here, we describe a method optimized to successfully hybridize RNA species, even of low abundance, in adult mouse bone and cartilage samples. This method involves tissue fixation with paraformaldehyde, demineralization with Morse's solution and paraffin embedding, all of which can be completed in 4 days. Sections are then generated and hybridized using a 1-day standard protocol. Sections prepared using this method are compatible with immunostaining and standard staining procedures for skeletal tissues.

Key words Bone, Cartilage, Gene expression, In vivo, RNA in situ, Tissue section

1 Introduction

The adult skeleton is mainly composed of bone and cartilage. These two distinct tissues are rich in extracellular matrix and sparse in cells. Their matrix consists of a fibrillar collagen network that entraps a dense gel of proteoglycans in the case of cartilage and that is mineralized in the case of bone. The activities of chondrocytes, osteoblasts, and osteoclasts are tightly controlled under homeostatic healthy conditions according to mechanical loading, hormonal influences and aging, and they can be drastically altered under pathological, genetic, and environmental conditions. Bone and cartilage diseases, namely, osteoarthritis and osteoporosis, affect a large subset of the adult and aging human population [1]. Research on these and other diseases often involves animal models. Assessing gene expression is key to analyzing cell activities in these models. This can be done using qRT-PCR and bulk sequencing of RNA extracted from tissues, using single-cell RNA

Tariq M. Haqqi and Véronique Lefebvre (eds.), *Chondrocytes: Methods and Protocols*, Methods in Molecular Biology, vol. 2245, https://doi.org/10.1007/978-1-0716-1119-7_6, © Springer Science+Business Media, LLC, part of Springer Nature 2021

sequencing, and using RNA in situ hybridization. The latter method involves hybridization of target RNAs in tissue sections with labeled antisense RNA probes. It very advantageously complements the others by allowing for visualization of gene expression in individual cells in morphologically preserved tissue. This is especially important for tissues that house several cell types and in which each cell type population is composed of cells at multiple differentiation stages or activity levels, such as bone, growth plate and articular cartilage.

Standard tissue processing and RNA in situ hybridization methods have proven to be challenging for adult bone and cartilage due to the low cellularity of the tissues, their dense extracellular matrix, and the mineralization of this matrix in the case of bone. Demineralization of skeletal samples is recommended for tissue, cellular, and molecular analysis of paraffin or frozen sections. This process typically involves incubation of the samples for extended periods of time in solutions at acidic pH or containing the cation chelator ethylenediaminetetraacetic acid (EDTA), increasing the chance of RNA degradation due to the release of RNases from possibly incompletely fixed tissue areas. In this chapter, we describe a quick and RNA-friendly method for sample processing and RNA in situ hybridization. Adult mouse skeletal tissues are fixed using 4% paraformaldehyde for one or two days, demineralized using Morse's solution for another one or two days, and embedded in paraffin (*see* **Note 1**). Sections are then made and subjected to a one-day RNA in situ hybridization assay (RNAscope technology, Advanced Cell Diagnostics). Thus, results can be obtained within a week. We show that this method fully preserves RNA integrity in both cartilage and bone tissues. It robustly detects mRNA targets of both high and low cellular abundance.

2 Materials

2.1 Reagents and Solutions

1. Diethyl pyrocarbonate (DEPC).
2. RNase-free water, commercially available or prepared in house using 0.1% DEPC. Add 1 ml of DEPC to 1000 ml MilliQ water. Stir vigorously until all DEPC globules disappear. Let sit at room temperature for 1 h to overnight. Autoclave for 45 min to fully inactivate DEPC.
3. RNaseZap™. This RNase decontamination solution is recommended to inactivate any RNase traces from glassware, plastic surfaces, countertops, and pipettes. Apply RNaseZap solution to the surfaces to be cleaned and then rinse with RNase-free water.
4. Phosphate buffered saline (PBS) prepared in RNase-free water.

5. 4% paraformaldehyde (PFA) diluted in RNase-free PBS from 16% stock.
6. Sodium citrate.
7. Formic acid.
8. Morse's solution (10% sodium citrate and 22.5% formic acid). Dissolve 100 g sodium citrate in 500 ml RNase-free water. Add 240 ml 94.5% formic acid stock (VWR, Cat. No. 97064-708). Add RNase-free water up to 1 l. Prepare no more than 3 days before use.
9. Xylene.
10. 100% ethanol.

2.2 Histology and Other Supplies

1. Dissection tools: scissors, forceps, and scalpels.
2. Insulin syringes and needles.
3. Tissue processing/embedding cassettes.
4. Screw-capped bottles with large opening (allowing the entry of embedding cassettes).
5. Thermometer.
6. Paraffin wax.
7. SuperFrost Plus slides.
8. ImmEdge™ hydrophobic barrier pen.
9. Tissue-Tek manual slide staining set, including white and green (xylene resistant) dishes, and two vertical 24-slide racks.
10. Square-mouth glass Wheaton staining jars (10-slide capacity).
11. VectaMount™ permanent mounting medium (Vector Laboratories, Cat. No., H-5000).
12. Cover glasses (24 × 50 mm).

2.3 Equipment

1. Microtome.
2. Water bath or incubator capable of holding temperature at 40 ± 1 °C.
3. Drying oven capable of holding temperature at 60 ± 1 °C.
4. Bright-field microscope and accessories.
5. Hybridization oven and slide tray set (Advance Cell Diagnostics [ACD] or equivalent).

2.4 RNAscope Reagents (ACD)

1. RNAscope 2.5 HD duplex detection kit (chromogenic) (for duplex detection), or RNAscope 2.5 HD Assay—RED (for single detection).
2. RNAscope probes for positive and negative controls and for RNAs of interest.

3. RNAscope hydrogen peroxide.
4. Wash buffer kit.
5. Custom pretreatment reagent.

3 Methods

3.1 Preparation of Tissue Sections

RNase-free conditions are not required for **steps 1**–**7**. Speed is the most important aspect in this part of the protocol. Proceed with one animal at a time and get the tissues of interest dissected out and immersed into fixative as quickly as possible (within 10 min), otherwise dying cells will release RNases, which will quickly destroy RNAs before the tissue is fully fixed.

1. Euthanize one mouse at a time via CO_2 asphyxiation followed by cervical dislocation. Immediately after death is confirmed, skin the mouse where appropriate and dissect out the desired skeletal elements. Cut off as much muscle and other nonskeletal tissues as possible and suitable.
2. Bore a few holes or make small slits in large bones and joints to ensure that the fixative and wash solutions rapidly access internal structures, such as bone marrow and synovial cavities.
3. Using an insulin syringe, inject small amounts of 4% PFA into remaining muscles to ensure better penetration of the fixative through the whole tissue.
4. Place the samples into tissue-embedding cassettes and submerge the cassettes in 4% PFA solution in a screw-capped bottle. Place the bottles on a rocker or rotator and incubate at room temperature for 24–48 h (depending on mouse age and tissue size).
5. Wash the samples for three times 15 min in RNase-free PBS.
6. Fill the bottles with Morse's solution (about 30 ml per sample). Allow samples to decalcify at room temperature for 24–48 h (*see* **Notes 2** and **3**).
7. Wash the samples for three times 15 min in RNase-free PBS.
8. Dehydrate samples in a graded ethanol series (70%, 80%, 90%, 95%, 100%, and 100%) followed by two changes in xylene, and two changes in paraffin. Incubate the samples in each solution for at least 1 h (*see* **Note 4**).
9. Embed the tissues in paraffin blocks using a standard method.
10. Cut 4 to 6-μm-thick tissue sections with a microtome (*see* **Note 5**). Float sections in a freshly cleaned DEPC-water bath at 40–45 °C until they have smoothed out on the water's surface (*see* **Note 6**).

11. Carefully lift the sections onto Superfrost Plus slides (*see* **Note 7**). Position the section as close to the middle of the slide as possible on both axes (*see* **Note 8**).
12. Dry the slides at 45 °C on a cleaned slide warmer until the sections have fully spread (about 2 h), and then dry them overnight at room temperature on a clean surface covered with 2–3 layers of Kimwipes in a protected location (e.g., an empty drawer or a cleaned slide dryer).
13. Store the slides in a slide box for up to 3 months at 4 °C (*see* **Notes 9** and **10**).

3.2 RNA In Situ Hybridization

The RNA in situ assay itself could in principle be carried out following any previously described protocol for radioactive or fluorescent probes [2–4]. We routinely use the RNAscope method according the manufacturer's instructions, but with the following adaptations:

1. Prepare all solutions in RNase-free water (*see* Subheading 2.1).
2. Replace the heated target retrieval and protease steps with an incubation of the slides with ACD Custom Pretreatment Reagent for 30 min at 40 °C. This solution is specifically designed for tissue sections that easily detach from the slides during target retrieval at high temperature (98–100 °C).

3.3 Image Acquisition and Data Analysis

Image the slides under regular bright-field microscopy conditions. Signals generated with the RNAscope method will appear as colored dots within cells counterstained with hematoxylin (light purple). Signal colors (purplish red, bluish green, etc.) will depend on the types of probes selected for the assay. Densitometry software (e.g., NIH ImageJ) can be used to quantify RNA signals.

Figure 1 illustrates the superiority of Morse's solution over EDTA solution to preserve RNA integrity during skeletal tissue demineralization.

Figure 2 compares RNA signals obtained for three marker genes for skeletal tissues: *Acan,* a gene highly and specifically expressed in chondrocytes (encoding the core protein of aggrecan); *Runx2*, a gene expressed at a low level in prehypertrophic and hypertrophic growth plate chondrocytes and in osteoblasts (encoding the RUNT-domain transcription factor 2); and *Dmp1*, a gene specifically and substantially expressed in growth plate terminal chondrocytes, osteoblasts, and osteocytes (encoding the dentin matrix protein 1).

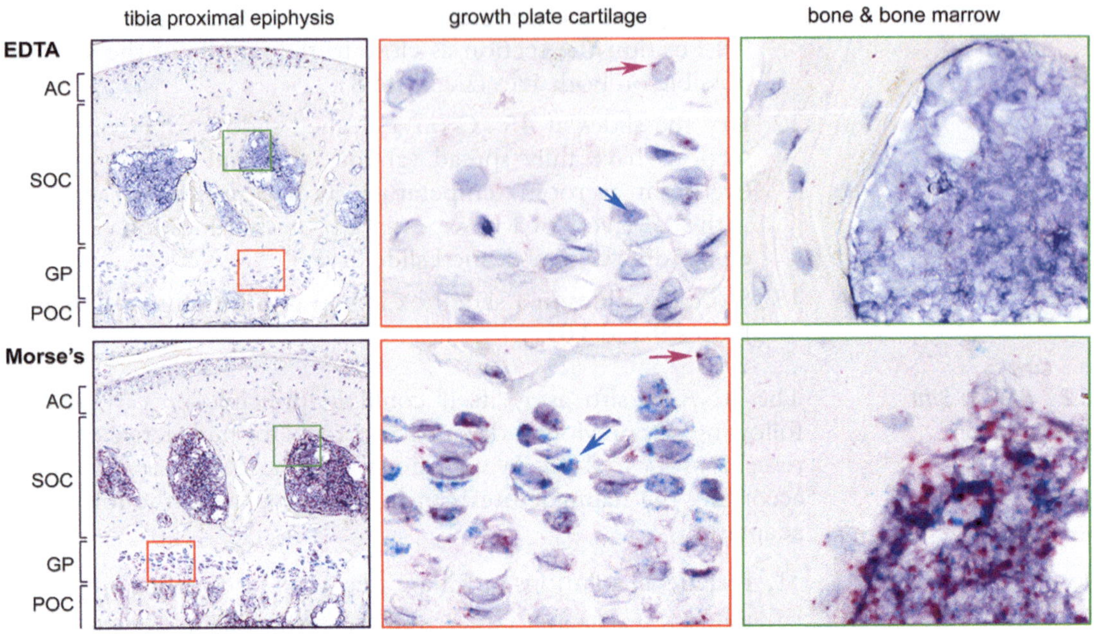

Fig. 1 Comparison of RNA signals obtained following tissue demineralization with 15% EDTA solution (pH 7.4) for 1 week (top panels) or with Morse's solution for 1 day (bottom panels). Parasagittal sections of 3-month-old mouse knees were hybridized with ACD positive control probes for *Ppib* RNA (bluish green signal) and *Polr2a* RNA (purplish red). *Ppib* (encoding peptidyl-prolyl cis-trans isomerase B) is expressed ubiquitously at a moderate-high level and *Polr2a* (encoding the largest subunit of RNA polymerase II) at a moderate-low level. Signals for *Polr2a* RNA are visible in only a few cells and those for *Ppib* RNA are visible in hardly any cell in EDTA-treated samples. Signals for both RNAs are much stronger and are visible in many more cells in Morse's solution-treated samples

4 Notes

1. This solution was first reported by A. Morse in 1945 [5, 6] to demineralize tooth specimens. It uses formic acid to powerfully decalcify tissues and sodium citrate to counteract formic acid-induced tissue swelling.
2. One day of fixation followed by 1 day of demineralization is sufficient for skeletal samples from mice up to 2 months of age. Two days for each step is recommended, but not required, for large samples from older mice.
3. If the step with Morse's solution is replaced with 1 week in 15% EDTA, pH 7.4, weak if any RNA signals are obtained. If the EDTA solution is reduced to 10% and is supplemented with 2% PFA, signals are as good as using Morse's solution, but this demineralization procedure can take more than 1 week for large bones.
4. We routinely perform these steps overnight using an automated tissue processor.

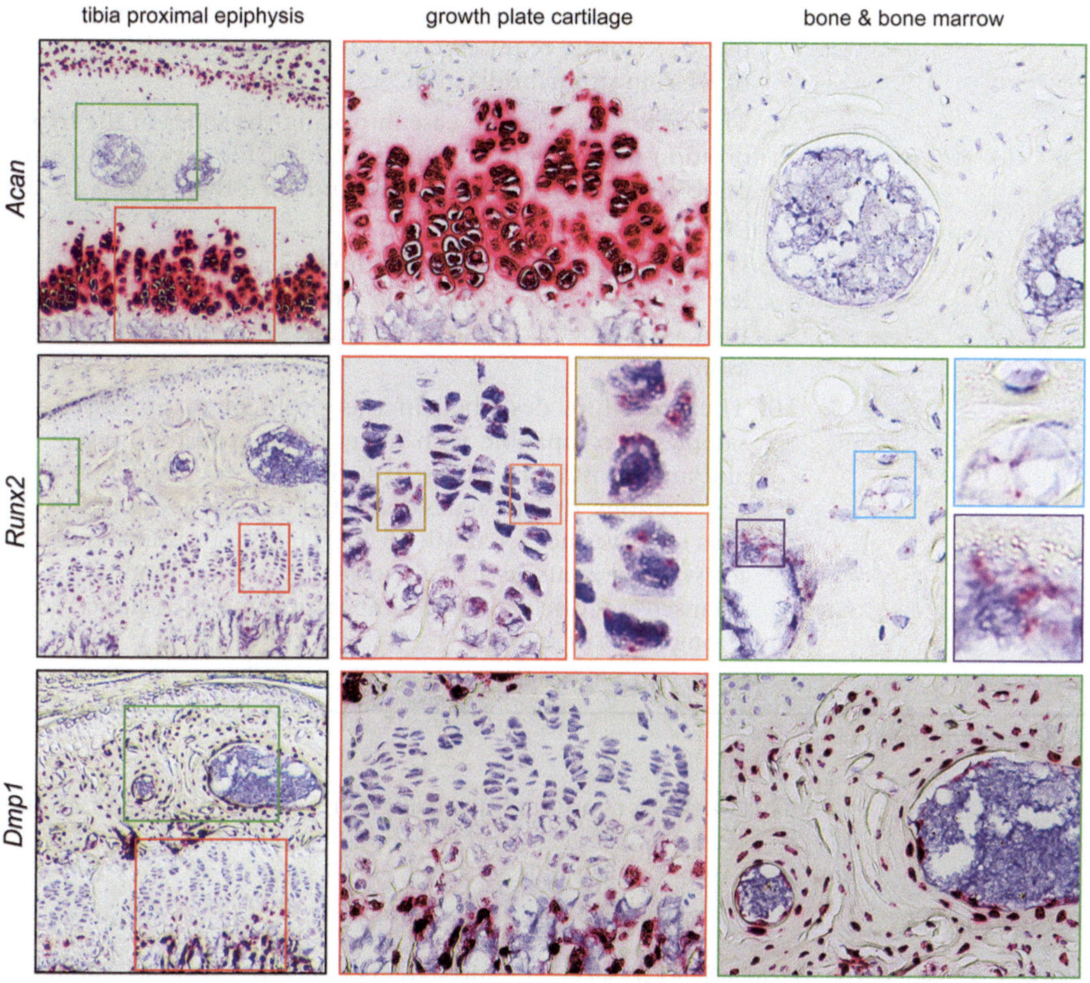

Fig. 2 Comparison of signals obtained for *Acan, Runx2*, and *Dmp1* RNAs. A 3-month-old mouse knee was fixed for 2 days and demineralized in Morse's solution for the next 2 days. Left panels, low-magnification images of parasagittal sections through the tibial proximal epiphysis, including articular cartilage (AC), secondary ossification center (SOC), growth plate cartilage (GP), and primary ossification center (POC). Middle and right panels, high-magnification images of regions boxed in the left panels. For *Runx2*, which shows very low expression, a few cells are presented at even higher magnification to show individual RNA signal dots. The signals for all three RNA probes are seen in purplish red. As expected, articular cartilage and growth plate cartilage show strong expression of *Acan*, but none in bone and bone marrow. *Runx2* is detected at a low level in growth plate chondrocytes and in bone and bone marrow cells. *Dmp1* is highly expressed in terminal chondrocytes of the growth plate, but not in less mature cells and articular chondrocytes. It is also strongly expressed in osteoblasts and osteocytes

5. Use RNase-Zap to clean the microtome stage surfaces, all other tools used for sectioning, and the water bath. Rinse and fill the water bath only with DEPC-treated water.
6. If sections begin to come apart too quickly, add room temperature water to lower the water bath temperature by a few degrees.

7. Superfrost Plus slides are specifically designed to firmly bind tissue sections, a requirement for the relatively harsh conditions of RNA in situ hybridization.
8. This location is best for ease in placing the slides in the hybridization tray without touching sections, for drawing a hydrophobic barrier around sections, and for other aspects of the protocol.
9. While sections can be stored for up to 3 months, it is nevertheless recommended to cut them from paraffin blocks no more than a week before performing the RNA in situ hybridization assay.
10. The procedure described in this protocol to process tissue samples is compatible with many types of histology staining and immunostaining assays. If such assays are desired on adjacent sections, dry the slides dedicated to these assays overnight on a slide warmer set at 40–45 °C. This is especially important for synovial joints, as articular cartilage surfaces tend to roll on themselves if dried at room temperature and then subjected to a long series of solvents.

Acknowledgments

This work was supported by NIH grant AR072649 and AR068308 to V.L.

References

1. Cross M, Smith E, Hoy D, Nolte S, Ackerman I, Fransen M, Bridgett L, Williams S, Guillemin F, Hill CL, Laslett LL, Jones G, Cicuttini F, Osborne R, Vos T, Buchbinder R, Woolf A, March L (2014) The global burden of hip and knee osteoarthritis: estimates from the global burden of disease 2010 study. Ann Rheum Dis 73(7):1323–1330. https://doi.org/10.1136/annrheumdis-2013-204763
2. Wilkinson DG (1995) RNA detection using non-radioactive in situ hybridization. Curr Opin Biotechnol 6(1):20–23. https://doi.org/10.1016/0958-1669(95)80004-2
3. Moorman AF, De Boer PA, Ruijter JM, Hagoort J, Franco D, Lamers WH (2000) Radio-isotopic in situ hybridization on tissue sections. Practical aspects and quantification. Methods Mol Biol 137:97–115. https://doi.org/10.1385/1-59259-066-7:97
4. Joeng KS, Regan J, Long F (2014) Radioactive in situ hybridization to detect gene expression in skeletal tissue sections. Methods Mol Biol 1130:217–232. https://doi.org/10.1007/978-1-62703-989-5_16
5. Morse A (1945) Formic acid-sodium citrate decalcification and butyl alcohol dehydration of teeth and bones for sectioning in paraffin. J Dent Res 24:143–153
6. Shibata Y, Fujita S, Takahashi H, Yamaguchi A, Koji T (2000) Assessment of decalcifying protocols for detection of specific RNA by non-radioactive in situ hybridization in calcified tissues. Histochem Cell Biol 113(3):153–159. https://doi.org/10.1007/s004180050434

Chapter 7

MicroRNA In Situ Hybridization in Paraffin-Embedded Human Articular Cartilage and Mouse Knee Joints

Helal Endisha and Mohit Kapoor

Abstract

MicroRNA (miRNA) in situ hybridization (ISH) is a highly sensitive method that allows for the detection of expression and distribution of miRNAs in fixed paraffin-embedded tissues. MiRNA ISH requires time-consuming optimization based on the tissue type analyzed, method of tissue fixation, and miRNA detection probe. Here, we provide the optimized miRNA ISH protocol for human cartilage and mouse whole knee joints that also entails the necessary steps for sample collection, processing, and preparation for high-quality ISH staining.

Key words Microrna, In situ hybridization,, Cartilage,, Mouse knee joint,, Locked nucleic acid,, Formalin-fixed, Paraffin-embedded (FFPE).

1 Introduction

MicroRNA (miRNA) in situ hybridization (ISH) is a technique that can be used for both the localization and detection of microRNA sequences at the cellular level in fixed tissue sections, providing crucial information regarding the spatial profile of miRNAs. Here we provide a detailed miRNA ISH protocol tailored specifically for miRNA detection in human articular cartilage and whole mouse knee joint sections with high sensitivity and specificity [1–4]. This protocol also includes all the steps required for sample collection, processing, and preparation for ISH. Following the steps for proper sample and reagents preparation is crucial for miRNA ISH to be successful. This protocol can be performed on formalin-fixed human tissue; however, we have described an alternative fixation protocol using paraformaldehyde (PFA) that can be applied to both human cartilage and mouse knee joints to achieve high-quality ISH staining. Specifically, for human cartilage sections, which contain subchondral bone and whole mouse knee joints; we recommend a gentle method of decalcification using 10% ethylenediaminetetraacetic acid (EDTA), which preserves the integrity of the tissue.

Tariq M. Haqqi and Véronique Lefebvre (eds.), *Chondrocytes: Methods and Protocols*, Methods in Molecular Biology, vol. 2245, https://doi.org/10.1007/978-1-0716-1119-7_7, © Springer Science+Business Media, LLC, part of Springer Nature 2021

After sample preparation, the miRNA ISH is a one-day protocol. Appropriate Proteinase-K treatment is required to demask miRNAs in order for the miRNA probe to hybridize to the miRNA sequence and varies based on tissue type and extent of fixation. Thus, we optimized the Proteinase-K concentration and duration of treatment for both human cartilage and mouse knee joints that exposed miRNA while maintaining tissue structure. The miRNA probe that we have used in our experiments contains a locked nucleic acid (LNA) modification, which improves probe sensitivity and specificity [5]. This reduces the likelihood of cross-hybridization to similar miRNA sequences, such as those in the same miRNA family. Since this is a chromogenic ISH protocol, the LNA miRNA probes contain a double-digoxigenin (DIG) label, to which an anti-DIG labeled antibody binds. The anti-DIG antibody is conjugated with an alkaline phosphatase (AP) enzyme that converts soluble 4-nitro-blue tetrazolium (NBT) and 5-bromo-4-chloro-3′-indolylphosphate (BCIP) substrates into NBT-BCIP, an insoluble blue precipitate, enabling miRNA localization to be visualized by routine bright-field microscopy. From our experience, previous nuclear counter stain procedures result in pale tissue staining, which makes microscope imaging challenging; thus nuclear counter staining was also optimized to improve contrast and tissue visualization, resulting in high-quality images clearly depicting miRNA localization.

2 Materials

The miRNA ISH protocol must be performed in a nuclease-free environment and precautionary measures must be taken to avoid RNase contamination of the reagents and miRNA degradation in the samples (*see* **Note 1**).

2.1 Sample Preparation

1. Sterile scissors.
2. Sterile scalpel.
3. Sterile forceps.
4. 1× phosphate-buffered saline (PBS) (sterile or autoclaved).
5. 15 mL and 50 mL canonical tubes.
6. RNase-free water.
7. Superfrost™ Plus slides.
8. 4% paraformaldehyde (PFA): In a fume hood, heat 225 mL RNase-free water to 55–57 °C and add 10 g PFA with constant stirring until the solution becomes clear (20–60 min). To help dissolve the PFA, add 1 M NaOH dropwise while stirring. Remove from heat, and add 25 mL 10× PBS. When solution

reaches room temperature, adjust pH to 7.4. Store at 4 °C for up to 1 month.

9. 10% EDTA: Add 50 g EDTA disodium dihydrate to 400 mL RNase-free water while stirring. Add approximately 3 mL of 5 M NaOH. Stir well and adjust pH to 7.4. Add RNase-free water for a final volume of 500 mL.
10. Rocking shaker.
11. 70% anhydrous ethanol.
12. Embedding cassettes.
13. JFC solution.
14. Paraffin wax.
15. Spin Tissue Processor (e.g., Model Microm STP-120, Thermo Scientific).

2.2 Deparaffinization

1. Autoclavable Coplin jars. To avoid potential cracking with glass Coplin jars, polypropylene Coplin jars (Sigma-Aldrich) are recommended.
2. 100%, 96%, and 70% anhydrous ethanol. Freshly prepare on day of use.
3. Xylene.
4. RNase-free water.
5. 1× sterile PBS.

2.3 In Situ Hybridization

1. Milli-Q, or other ultrapure, water.
2. 1.5 mL microcentrifuge tubes.
3. Sterile filter tips.
4. Benchtop microcentrifuge.
5. Proteinase-K: 200 μg/mL.
6. Proteinase-K buffer: To 900 mL RNase-free water add 5 mL 1 M Tris–HCl (pH 7.4), 2 mL 0.5 M EDTA, and 2 mL 5 M NaCl. Adjust volume to 1 L and autoclave.
7. DIG-labeled LNA miRNA probe of interest: 25 μM stock concentration (*see* **Note 2**).
8. Double-DIG labeled LNA scramble miRNA probe (negative control): 25 μM stock concentration.
9. 5′ DIG-labeled LNA U6 snRNA probe (positive control): 0.5 μM stock concentration.
10. Formamide-free miRNA ISH buffer (2×) (Qiagen).
11. Parafilm.
12. Block heater.
13. Autoclaved coverslips.

14. 45° angled up coverslip forceps (e.g., 0.1 × 0.08 mm; Dumont #5/45, Fine Science Tools).
15. Slide denaturization and hybridization system (e.g., Thermo-Brite® Model S500, Leica).
16. Molecular biology grade (ultrapure) 20× saline sodium citrate (SSC) buffer: Prepare 5× SSC (250 mL 20× SSC + 750 mL water), 1x SSC (50 mL 20× SSC + 950 mL water), and 0.2 × SSC (10 mL 20× SSC + 950 mL water). Autoclave all solutions.
17. Hydrophobic pen.
18. Humidity chamber (e.g., StainTray slide staining system, Sigma-Aldrich).
19. Bovine serum albumin (BSA) (lyophilized powder): Make a 30% BSA solution by dissolving 3 g BSA in 10 mL sterile 1× PBS in a 50 mL tube.
20. Antibody blocking solution: phosphate-buffered saline (PBS), 0.1% Tween 20, 2% sheep serum, 1% BSA.
21. Antibody diluent solution: PBS, 0.05% Tween 20, 1% sheep serum, 1% BSA.
22. Sheep anti-DIG-AP antibody.
23. PBS-Tween (0.1%), pH 7.4: Add 1 mL Tween 20 to 1 L PBS and autoclave.
24. NBT/BCIP ready-to-use tablets.
25. Levamisole (100 mM).
26. KTBT (AP stop solution): 50 mM Tris–HCl, 150 mM NaCl, 10 mM KCl. To 900 mL RNase-free water, add 7.9 g Tris–HCl, 8.7 g NaCl, 0.75 g KCl, adjust volume to 1 L and autoclave. No need to adjust pH.
27. Nuclear Fast Red™ counter stain (Vector laboratories).
28. 20 mL syringe and 0.22 μm filter.
29. Mounting medium; we recommend Eukitt® (Sigma-Aldrich), a fast-drying, xylene-based mounting medium ideal for pale color preparations.

3 Methods

3.1 Sample Preparation

1. (a) Mouse knee joint: Prior to incision, sterilize the hind legs with 70% ethanol and isolate mouse knee joints by cutting the tibia below the knee joint and the femur close to the hip joint using sharp sterile scissors (*see* **Note 3**).

   (b) Human cartilage: With research ethics board approval, obtain human knee cartilage from knee joints of patients undergoing arthroplasty. Store cartilage specimens in a 50 mL conical tube with ice-cold, sterile 1× PBS at 4 °C. Under a biosafety cabinet, cut cartilage into 5–6 mm sections using a sterile scalpel.

2. Tissue fixation: Using sterile forceps, place mouse knee joint or human cartilage in a 15 mL conical tube filled with 10 mL 4% PFA. Fix for 48 h at 4 °C.
3. Wash 3 × 10 min with RNase-free water on a rocking shaker.
4. Decalcification: Transfer tissue to a 50 mL conical tube and add 40 mL 10% EDTA. Keep at 4 °C for 2–3 weeks, changing the EDTA solution each week (*see* **Note 4** for determining when tissue has been decalcified).
5. Wash 3 × 10 min with RNase-free water on a rocking shaker.
6. Place tissue in an embedding cassette and immerse in PBS.
7. Use a spin tissue processor (e.g., Microm STP-120, Thermo Scientific) to dehydrate, clear, and infiltrate the tissue with paraffin according to the following steps: (1) 2 h in 70% ethanol at 37 °C; (2) 1 h in 80% ethanol at 37 °C; (3) 2 h in 100% ethanol at 37 °C; (4) JFC solution for 15 h at room temperature; (5) heating dish for 1 h at 45 °C; (6) first paraffin dish for 1 h at 62 °C; (7) second paraffin dish for 1 h at 62 °C; (8) third paraffin dish for 1.5 h at 62 °C.
8. Remove cassettes from the dish and start to embed.
9. *Optional stop point*: store paraffin-embedded samples under RNase-free conditions at 4 °C until cutting.
10. Cut tissue sections at a thickness of 4 μm using a microtome and place paraffin ribbons in a water bath filled with RNase-free water heated to 45 °C.
11. Mount paraffin sections onto sterile Superfrost™ Plus slides.
12. Allow paraffin sections to dry 1–2 h at room temperature and store at 4 °C before conducting ISH. Slides can be stored at 2–8 °C for up to 1 week. If ISH is to be performed greater than a week after paraffin embedding, cut new sections from the paraffin blocks (which are to be stored at 4 °C) in the week of performing ISH.
13. The day before performing ISH, place slides, section side up, in a 60 °C oven for 45 min to melt the paraffin. Store slides in a slide box overnight at 4 °C.

3.2 Deparaffinization

Please note that this is a one-day protocol (approx. 7 h) that must be completed in 1 day.

1. Deparaffinize slides in Coplin jars of xylene and decreasing concentrations of ethanol as follows: 3 Coplin jars of xylene (5 min each), 3 Coplin jars of 100% ethanol (immerse up and

down ten times in each of the first two to rinse; 5 min in third jar), 2 Coplin jars of 96% ethanol (immerse up and down ten times in first jar, 5 min in second jar), 2 Coplin jars of 70% ethanol (immerse up and down ten times in first jar, 5 min in second jar), and 1 Coplin jar of 1× PBS (5 min) (*see* **Note 5**).

Tip: Sterilize a humidified staining chamber by spraying with 70% ethanol then fill reservoir with RNase-free water and place in a cell culture incubator during the deparaffinization step in order to reach 37 °C for the following step.

3.3 In-Situ Hybridization

1. Immediately before use, dilute Proteinase-K stock solution (200 μg/mL) with the previously prepared Proteinase-K buffer to the appropriate concentration (*see* **Note 6**). Apply enough Proteinase-K to fully cover tissue sections (approx. 100 μL per human cartilage section and 300 μL per mouse knee joint section) and incubate slides in a humidified staining chamber for 10–20 min at 37 °C (in a cell culture incubator) (*see* **Note 6**).
2. While slides are incubating, prepare LNA miRNA probe hybridization mix (*see* **Note** 7). Dilute the 2× miRNA ISH buffer to 1× with RNase-free water. Place the desired volume of LNA probe in a 1.5 mL sterile microcentrifuge tube. Denature the probes in a block heater at 90 °C for 4 min. Cover the cap with parafilm to provide a secure seal and minimize evaporation. Immediately place the tubes on ice followed by a short spin down in a benchtop microcentrifuge for 30 s. Add the 1× miRNA ISH buffer to each tube depending on the desired final probe concentration (*see* **Note** 7). Keep diluted probes at room temperature until they are added onto the tissue sections.
3. After Proteinase-K incubation, wash slides twice, 5 min each wash, in a Coplin jar with sterile 1× PBS.
4. Carefully tap slides to remove excess PBS off tissue sections and dab with a Kimwipe, if necessary, taking care to avoid touching the tissue sections.
5. Apply 50 μL of hybridization mix to tissue sections and use sterile forceps to cover the sections with a sterile coverslip. Take care to avoid air bubbles. Dampen the humidifying strips of the hybridization system with RNase-free water. Insert the slides in the hybridization system and set a program for 1 h at 55 °C. Note that individual probes require optimization of hybridization temperatures (*see* **Note 8**). If a hybridization system is not available, slides can be placed in a humidified chamber in an oven set at 55 °C (*see* **Note 9**).
6. During hybridization (**step 5**), fill 2 Coplin jars with 50 mL 5× SSC, 2 Coplin jars with 50 mL 1× SSC, and 3 Coplin jars with

50 mL 0.2× SSC. Keep one 5× SSC Coplin jar and one 0.2× SSC Coplin jar at room temperature. Place the remaining Coplin jars in a heated water bath set to the hybridization temperature.

7. After hybridization, transfer the slides to the Coplin jar containing 5× SSC for 5 min at room temperature. Note: the coverslips will easily detach and slide off the slides to allow for subsequent washes. Using a thermometer, check that the Coplin jars containing the SSC buffers in the water bath have reached the hybridization temperature (55 °C). For stringent washes, use sterile forceps to transfer the slides to the warmed Coplin jars in the following order (*see* **Note 10**): 5× SSC (5 min), 2 Coplin jars with 1× SSC (5 min each), 2 Coplin jars with 0.2× SSC (5 min each). Finally, transfer slides to a Coplin jar with 0.2× SSC at room temperature for 5 min.
8. Set the oven to 30 °C for **step 13**.
9. Transfer slides to a Coplin jar with 1× PBS. Remove slides from PBS and carefully blot any excess liquid with a Kimwipe while taking care to avoid touching the tissue. Carefully, apply a hydrophobic barrier around tissue sections using a hydrophobic pen. All subsequent steps are to be conducted in a humidifying chamber ensuring slides do not dry out.
10. Incubate slides with 50 μL antibody blocking solution for 15 min at room temperature in a humidifying chamber (*see* **Note 11**).
11. Dilute sheep anti-DIG-AP 1:500 in antibody diluent solution (*see* **Note 12**). Tap off blocking solution and apply 50 μL sheep anti-DIG-AP on the tissue section. Incubate for 60 min at room temperature in a humidifying chamber (*see* **Note 13**).
12. Wash the slides in three Coplin jars of PBS–Tween, 3 min each wash.
13. Prepare AP substrate immediately before use: in a 15 mL tube add 1 NBT-BCIP tablet to 10 mL Milli-Q water and add 20 μL levamisole. Wrap the tube in aluminium foil to protect from light. Add 50 μL of AP substrate to the sections and place slides in a covered humidifying chamber (protected from light). Incubate in an oven at 30 °C for 2 h (*see* **Note 14**).
14. Fill 2 Coplin jars with 50 mL autoclaved KTBT buffer (AP stop solution). Incubate slides in KTBT buffer twice, 5 min each incubation.
15. Wash slides in 2 Coplin jars of Milli-Q water, 1 min each wash.
16. Add 100 μL of filtered Nuclear Fast Red™ on tissue sections and incubate for 10 min.

Tip: Filter Nuclear Fast Red™ counter stain using a 0.22 μm syringe filter attached to a 20 mL syringe to remove color precipitate.

17. Carefully tap off Nuclear Fast Red™ and wash in a Coplin jar containing tap water for 5 min.
18. Dehydrate the slides in Coplin jars containing increasing concentrations of ethanol: 2 Coplin jars of 70% ethanol (gently immerse up and down ten times then 1 min incubation), 2 Coplin jars of 96% ethanol (gently immerse up and down ten times followed by 1 min incubation), 2 Coplin jars of 100% ethanol (gently immerse up and down ten times followed by 1 min incubation).
19. Remove the slides from ethanol, directly add 2 drops of Eukitt® mounting medium using a plastic transfer pipette, and cover with a coverslip.

    Tip: Keeping the slide wet with ethanol will help the mounting medium disperse across the slide without air bubbles.
20. Let the blue precipitate settle and the mounting medium dry overnight before imaging using a bright field microscope (*see* **Note 15**).

4 Notes

1. Ensure all surfaces for RNA work are cleaned with an RNase decontamination solution, such as RNaseZap or RNase AWAY. Gloves must be worn at all times. Prepare all buffers and reagents with RNase-free water or autoclaved 0.1% diethylpyrocarbonate (DEPC)-treated water. All glassware and coverslips should be wrapped in aluminum foil and autoclaved. Failure to maintain RNase-free conditions during tissue processing and sectioning can result in little-to-no miRNA detection.
2. To avoid multiple freeze–thaw cycles, aliquot all probes into 5 μL aliquots in RNase-free tubes and store at −20 °C. For short-term use, probes can be stored at 4 °C for up to 4 weeks.
3. For mouse knee joints, remove as much muscle and surrounding connective tissues as possible to ensure the fixation solution penetrates effectively. However, care must be taken to (1) avoid twisting or dislocating the knee joint while removing the surrounding tissue and (2) preserve the synovial tissue surrounding the knee joint.
4. In order to ensure release of calcium ions from the bone, tissue needs to be completely submersed in EDTA solution. We recommend 30–40 times the volume of the tissue. Complete decalcification of the mouse knee joint can be tested by slicing

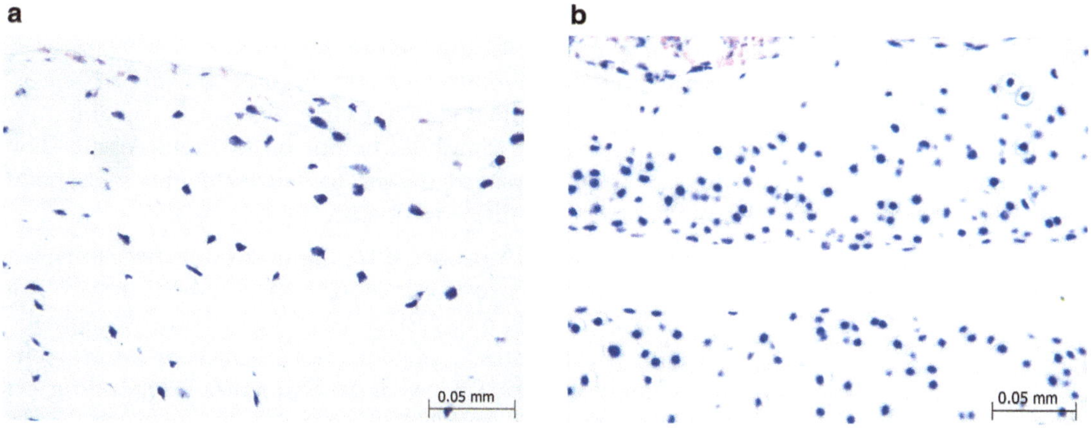

Fig. 1 Positive control staining of human cartilage and mouse knee joint sections using an LNA U6 snRNA probe (1 nM). Representative images of (**a**) human cartilage and (**b**) mouse knee joint (top: femoral condyle cartilage, bottom: tibial plateau cartilage) sections stained with 1 nM LNA U6 snRNA probe (positive control) and counterstained with Nuclear Fast Red™. Images (40×) taken using a bright-field microscope (Leica Microsystems DM750)

the distal portion of the bone (tibia or femur) using a sterile scalpel. The bone should have a rubbery consistency and the scalpel should slice easily through the bone with no resistance. Use this decalcification test after 2 weeks. If bone was not decalcified fully, keep in fresh 10% EDTA for 1 more week at 4 °C.

5. Dilute anhydrous ethanol in RNase-free water and use sterile forceps to transfer slides from one autoclaved Coplin jar to the next. Approximately 50 mL is sufficient to fill a Coplin jar and each Coplin jar can hold 5 slides vertically or 10 slides back-to-back.

6. Stronger fixation treatments, such as formalin, requires more concentrated and/or longer duration of Proteinase-K treatment. However, care must be taken to avoid excessive treatment, which can damage the tissue. Thus, the appropriate Proteinase-K treatment range must be empirically determined by testing the LNA U6 snRNA probe (positive control [Fig. 1]; *see* **Note 7** for probe preparation) using varying concentrations (5–20 μg/mL for human tissue; 0.5–5 μg/mL for mouse tissue) or durations (5–30 min) of Proteinase-K treatment at 37 °C. Once the optimum concentration and duration of Proteinase-K treatment has been established using the LNA U6 probe (i.e., most prominent tissue labeling with minimal tissue damage), the double-DIG labeled miRNA probe of interest can be tested using the same conditions. If a strong signal has been obtained with the LNA U6 probe, but not with your miRNA probe of interest, Proteinase-K treatment conditions must be optimized. *For human cartilage sections, we*

determined an optimum treatment of 20 μg/mL Proteinase-K for 20 min at 37 °C. For mouse knee joints, a Proteinase-K concentration of 7.5 μg/mL for 10 min at 37 °C worked best.

Tip: If a humidified staining chamber is not available, use an empty filter tip box and fill the bottom with wet tissue. The slides can then be placed on the top surface. This will create humidity when the filter tip box is closed.

7. For LNA U6 snRNA probe (0.5 μM stock concentration), we recommend a final probe concentration of 1 nM; however, a range of 0.1–2 nM can be tested. For your LNA miRNA probe of interest (25 μM stock concentration), a final probe concentration range of 40–100 nM is recommended depending on the abundancy of the miRNA, but ideal concentrations should be determined for each probe. Keep in mind that higher concentrations of the miRNA LNA probe may result in nonspecific extracellular matrix staining. A diluted probe volume of 40–50 μL is sufficient to cover the tissue. *For example, for a 1 nM final probe concentration of 0.5 μM LNA U6 snRNA probe dilute the stock 1:500 in 1× miRNA ISH buffer. For a 100 nM final probe concentration of 25 μM (stock concentration) LNA miRNA probe dilute the stock 1:250 in 1× miRNA ISH buffer.* Once optimum probe concentration has been determined, the remainder diluted probe can be aliquoted and stored at −20 °C. For reuse, thaw diluted probe at room temperature and apply directly on sections.

   Tip 1: During positive control optimization, it is recommended to stain sections from several paraffin blocks. If no signal is obtained with the positive control probe (LNA U6 snRNA), discard and reprepare all the reagents ensuring RNase-free conditions were maintained.

   Tip 2: If a strong signal was obtained with the LNA U6 probe and not the miRNA LNA probe and Proteinase-K conditions can not be modified (e.g., tissue becomes too fragile), try increasing concentrations of the miRNA LNA probe that will give a signal while avoiding nonspecific staining.

8. Hybridization temperatures influence the signal-to-noise ratio of LNA miRNA probes. Recommended temperature for probe hybridization is 30 °C below the RNA probe melting temperature (T_m). For most probes, the optimal temperature range for probe binding is typically between 50 and 60 °C. If the temperature is too low, the risk of unwanted cross-hybridization to similar sequences increases. However, a hybridization temperature of 60 °C can reduce probe binding capacity.

9. If a hybridization system is not available, use a humidified chamber (as suggested in **Note 6**) in an oven set to the hybridization temperature. However, in this case coverslips should be sealed on slides with nail polish.

10. To avoid nonspecific staining, ensure that SSC wash buffers are maintained at the hybridization temperature. High background signal of the LNA miRNA probe can be reduced by increasing the duration of the stringency washes.
11. Antibody blocking solution (2% sheep serum, 1% BSA in PBS–Tween): In a 15 mL canonical tube, add 300 μL sheep serum to 15 mL PBS–Tween. Remove 5 mL and place in a separate 15 mL canonical tube for the antibody diluent solution. To the remaining ~10 mL, add 330 μL 30% BSA.
12. Antibody diluent solution: To the 5 mL solution reserved from the blocking solution, add 5 mL 1× PBS and 330 μL 30% BSA to give a final concentration of 0.05% Tween, 1% sheep serum, and 1% BSA.
13. A range of 1:500 to 1:2000 sheep anti-DIG-AP antibody in antibody diluent solution is recommended. To strengthen a weak signal, replace the sheep anti-DIG-AP solution half-way through the incubation with fresh sheep anti-DIG-AP for the remaining incubation time.
14. For anticipated weak signals, AP substrate can be replaced half-way through incubation (i.e., after 1 h) with fresh AP substrate for the remaining hour. If a humidifying chamber is not available, wrap the humidified tip box (*see* **Note 6**) with aluminum foil.
15. The more abundant the target miRNA, the stronger the corresponding signal, resulting in a deep blue color precipitate.

Acknowledgments

The authors would like to thank Dr. Starlee Lively for her help with the review of this book chapter.

Funding: This work is supported, in part, by the Canada Research Chairs Program and The Natural Sciences and Engineering Research Council of Canada (NSERC).

References

1. Nuovo GJ (2008) In situ detection of precursor and mature microRNAs in paraffin embedded, formalin fixed tissues and cell preparations. Methods 44(1):39–46
2. Obernosterer G, Martinez J, Alenius M (2007) Locked nucleic acid-based in situ detection of microRNAs in mouse tissue sections. Nat Protoc 2(6):1508–1514
3. Jorgensen S et al (2010) Robust one-day in situ hybridization protocol for detection of microRNAs in paraffin samples using LNA probes. Methods 52(4):375–381
4. Nakamura A et al (2019) microRNA-181a-5p antisense oligonucleotides attenuate osteoarthritis in facet and knee joints. Ann Rheum Dis 78(1):111–121
5. Kloosterman WP et al (2006) In situ detection of miRNAs in animal embryos using LNA-modified oligonucleotide probes. Nat Methods 3(1):27–29

Chapter 8

Laser Capture Microdissection of Mouse Growth Plate Cartilage

Bijal Kikani and Julian C. Lui

Abstract

The ability to identify, isolate, and study pure populations of cells is critical for understanding normal physiology in organs and tissues, which involves spatial regulation of signaling pathways and interactions between cells with different functions, expression profiles, and lineages. Here, we focus on assessing the growth plate cartilage, composed of multiple functionally and histologically distinct zones, to investigate temporally and spatially dependent gene expression differences. In this chapter, we describe the method of laser capture microdissection to isolate chondrocytes from different zones of differentiation in the mouse growth plate cartilage for RNA isolation, and subsequent downstream applications, such as RNA-sequencing and quantitative real-time PCR. We also provide an assessment of different factors contributing to the integrity of the isolated RNA, such as staining methods and procedures in RNA isolation.

Key words Laser capture microdissection, Growth plate, Cartilage, Bone, Mouse

1 Introduction

Postnatally, linear bone growth occurs in the epiphysis near the ends of long bones in a cartilaginous structure known as the growth plate. The growth plate cartilage is primarily composed of chondrocytes arranged in three spatially, morphologically, and functionally distinct zones: resting, proliferative, and hypertrophic zones [1, 2]. Chondrocytes in the resting zone act as progenitor cells that can self-renew and give rise to chondrocytes in the proliferative zone [3, 4]. Chondrocytes in the proliferative zone are then arranged in columns parallel to the long axis of the bone and undergo rapid cell proliferation. Toward the bottom of the column, chondrocytes stop dividing and begin to undergo hypertrophy. Eventually, the last few layers of hypertrophic chondrocytes start to undergo matrix calcification and trigger vascularization. The chondrocytes either become apoptotic or undergo transdifferentiation to become osteoblasts [5, 6], and cartilage matrix is replaced

Tariq M. Haqqi and Véronique Lefebvre (eds.), *Chondrocytes: Methods and Protocols*, Methods in Molecular Biology, vol. 2245, https://doi.org/10.1007/978-1-0716-1119-7_8, © Springer Science+Business Media, LLC, part of Springer Nature 2021

by bone matrix in the metaphysis. The spatial progression of progenitor cells to transit-amplifying cells to terminally differentiated cells in the growth plate is coordinated both by intracellular mechanisms, such as expression of SOX9 and other transcription factors [7], and by various paracrine signaling pathways, including the PTHrP-IHH feedback loop [1], BMP [8] and WNT signaling gradients [9]. Therefore, the ability to precisely dissect out chondrocytes from different spatial locations of the growth plate is important for studying their unique expression profile and to improve our fundamental understanding of the growth plate.

Laser capture microdissection (LCM) is a technique that facilitates this study by targeted extraction of cell populations [10, 11] or single cells for downstream analysis, such as reverse transcription of RNA for quantitative real-time PCR [12–14]. The Arcturus ^{XT} LCM technique is an automated way to extract pure cell populations from frozen or paraffin embedded tissue sections without altering cellular morphology (ThermoFisher). To perform the dissection, this system uses infrared and ultraviolet laser beams to capture and precisely cut out targeted cell populations from histological sections. Once the user marks the cell population of interest, the instrument places a CapSure LCM cap containing a transparent ethylene vinyl acetate layer [15] directly on top of the marked region (ThermoFisher). An infrared laser beam is then activated by the user, which causes the thermoplastic film on the cap to transiently melt and adhere to the indicated cluster of cells [15]. The ultraviolet beam is then activated to cut out the targeted cell population and the cap is lifted off the tissue. Following microdissection, RNA can be extracted from the tissue for cDNA synthesis and quantitative real-time PCR. This chapter delineates both the preparatory stages and the details of LCM on frozen tissue samples, including slide pretreatment methods, tissue sectioning, slide staining, and laser capture microdissection of mouse growth plate cartilage [16, 17]. We also compared different staining methods and RNA isolation reagents to assess their impact on RNA integrity.

2 Materials

2.1 Slide Pretreatment

1. Arcturus PEN membrane frame slides (Thermo Fisher Scientific).
2. 0.1% poly-L-lysine (Sigma-Aldrich).
3. Tissue-Tek Slide Staining Dish and Staining Rack (Sakura Finetek USA).
4. Chloroform.

2.2 Tissue Sectioning

1. Leica Cryostat CM3050S.
2. Accu-Edge Low Profile Blades.
3. Leica chuck.
4. Tissue embedding mold (Polysciences).
5. Optimum Cutting Temperature (OCT) Compound.
6. Brush to maneuver the section across the stage.
7. Poly-L-lysine–treated PEN membrane frame slides.
8. Frame supporting glass slide.
9. Dry ice.

2.3 Staining

1. 50 ml conical tubes.
2. Ultrapure water.
3. 50%, 70%, 95%, and 100% RNase-free ethanol.
4. Xylene.
5. Tissue-Tek Slide Staining Dish (solvent-resistant green dish) and Staining Rack.
6. Eosin Y (Sigma-Aldrich).
7. Mayer's Hematoxylin (Sigma-Aldrich).
8. Cresyl Violet acetate (Sigma-Aldrich).
9. Desiccator capsules.
10. Slide box.
11. Forceps to move slides.

2.4 LCM

1. Superfrost Plus/Colorfrost Plus Microslides, 75 × 25 mm.
2. 0.5 ml Gene-Amp Thin-Walled Reaction tube, Domed Cap.
3. HS LCM Caps Arcturus CapSure.
4. RLT Buffer from QIAGEN RNeasy mini kit.
5. Beta-mercaptoethanol (BME) to denature intracellular RNases that are released during lysis.
6. Dry ice.

2.5 RNA Isolation and Analysis

1. RNeasy Micro Kit (QIAGEN).
2. PicoPure RNA isolation kit (Arcturus).
3. Agilent 2100 Bioanalyzer (Agilent).
4. RNA Pico Chips (Agilent).

3 Methods

3.1 Poly-L-Lysine Slide Coating

Poly-L-lysine solution is used as an adhesive to prevent tissue detachment that may occur during slide staining. This solution has a strong polycationic nature that interacts favorably with anionic sites in the tissue samples (Sigma-Aldrich) (*see* **Note 1**).

1. Rinse Tissue-Tek slide staining dish and staining rack with chloroform to keep materials RNase free. Remove chloroform and air dry the dish and rack briefly in a chemical hood.
2. Prepare 0.01% poly-L-lysine solution by diluting in UltraPure water. Pour the 0.01% poly-L-lysine solution into the staining dish.
3. Place PEN membrane frame slides in the staining rack.
4. Place the staining rack in the staining dish to submerge the PEN membrane frame slides in 0.01% poly-L-lysine solution for 5 min.
5. Take the staining rack and slides out of the staining dish.
6. Allow slides to air dry overnight at room temperature by moving the staining rack into an empty staining dish.

3.2 CryoStat Sectioning

CryoStat is used to create 12 μm sections of frozen tissue from 1- to 4-week-old mouse bones (*see* **Note 2**).

1. Add enough OCT compound into a tissue embedding mold to cover the bottom surface. Place the tissue embedding mold on ice.
2. Briefly immerse freshly dissected mouse tibial or femoral epiphyses in OCT compound in a petri dish at room temperature.
3. Use forceps to transfer the epiphysis into the tissue embedding mold with OCT compound on ice, and position the epiphysis in the center of the tissue embedding mold, with the sagittal plane facing down.
4. Transfer the tissue embedding mold with the epiphysis and OCT compound to dry ice.
5. When the OCT compound starts solidifying, add more OCT compound into the mold.
6. Allow the OCT compound in the mold to freeze completely. The tissue block can now be stored at −70 °C (*see* **Note 3**).
7. On the day of sectioning, the Cryostat should be turned on and allowed sufficient time for the machine to equilibrate to −20 °C. A range of −18 °C to −20 °C has been tested to work equally well on making sections of the mouse epiphysis at this age.

8. Remove the tissue block from the tissue embedding mold. To a Cryostat chuck, apply some OCT compound and place the frozen tissue block directly on the OCT compound. Place the Cryostat chuck in the Cryostat chamber. The OCT compound will freeze, setting the tissue block firmly mounted to the Cryostat chuck.
9. Insert the stem of the Cryostat chuck into the holder, orient the tissue as desired and tighten the chuck using the clamping screw.
10. Install a new Accu-Edge Low Profile Blade in the Cryostat.
11. Begin by resetting the tissue block all the way back, away from the blade. Use the stage forward button to bring the tissue block gradually closer to the blade. Carefully adjust the plane of your specimen (which determines the angle by which the blade hits the tissue block) using the levers in the Cryostat as needed.
12. Begin by cutting away at least 200 μm of the tissue block to advance the tissue block closer to the center portion of the epiphysis.
13. Make 12 μm sagittal sections of the epiphysis. Use the antiroll plate to help prevent sections from curling. Gently place the section on Superfrost plus slides. Stain the section briefly in eosin Y to check the location of the epiphysis where the sections were made. Chondrocyte columns and the secondary ossification center may not be visible if sections were made close to the edge of the epiphysis. Continue to make fine adjustments as needed.
14. When the desired location is reached for collecting tissue sections, make 12 μm sagittal sections on the epiphysis. A brush may be used to help prevent curling and to flatten the section.
15. Place a precoated poly-L-lysine PEN membrane slide on a frame supporting glass slide (*see* **Note 4**) (Fig. 1). Carefully

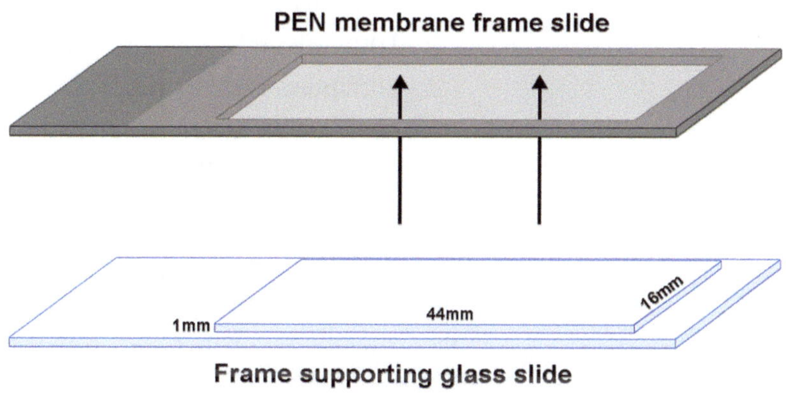

Fig. 1 Schematic diagram showing assembly of the frame supporting glass slide with PEN membrane frame slide for CryoStat Sectioning

collect the section by gently placing the PEN membrane frame slide (still in touch with the frame supporting glass slide) on the sectioned tissue. Because the PEN membrane frame slide is warmer than the sectioned tissue, the tissue will thaw and stick to the PEN membrane frame slide.

16. Repeat cutting and placing the sectioned tissue on the same PEN membrane frame slide. We suggest placing four sectioned tissues on each PEN membrane frame slide (*see* **Note 5**).
17. Freeze the PEN membrane frame slide in a slide box placed on dry ice.
18. Continue to make sections of the same sample on another PEN membrane slide. For each sample, we suggest making about 10 slides (four sections each).
19. The frozen sections should be stored at −80 °C. Staining and laser capture microdissection should be completed within 24–48 h to minimize RNA degradation (*see* **Note 6**).

3.3 Frozen Section Staining Methods

Note that frozen sections can be stained using either hematoxylin and eosin or cresyl violet.

3.3.1 Frozen Section H&E Staining

Hematoxylin is a basic dye that stains nucleic acids blue. Eosin, a counterstain for hematoxylin, is an acidic dye that stains proteins nonspecifically. This stain helps users visualize cell morphology.

1. Prepare 50 ml volumes of the following solutions in eleven separate 50 ml conical tubes:
   (a) Hematoxylin.
   (b) Eosin Y.
   (c) Bluing reagent.
   (d) 70% ethanol (2 tubes).
   (e) 95% ethanol.
   (f) 100% ethanol.
   (g) Ultrapure water (3 tubes).
2. Prepare 250 ml Xylene in a Tissue-Tek Slide Staining Dish (solvent-resistant green dish) and Staining Rack in a chemical hood.
3. Cool a slide box on dry ice. Immediately before staining, obtain frozen sectioned slides from −70 °C and transfer to the cooled slide box on dry ice.
4. Use forceps to dehydrate and stain slides individually, as follows:
   (a) 70% ethanol, 30 s.

(b) Ultrapure water, 30 s, use forceps to move slide up and down to remove OCT (*see* **Note 7**).

(c) Hematoxylin, 30 s.

(d) Ultrapure water, 15 s.

(e) (optional) Bluing reagent, 15 s; ultrapure water, 15 s.

(f) 70% ethanol, 15 s.

(g) Eosin Y, <1 s.

(h) 95% ethanol, 30 s.

(i) 100% ethanol 30 s.

(j) Xylene, 5 min or hold.

Perform the last step in a chemical hood, keep slides in xylene and start staining the next slide. Keep all slides in xylene until all slides are dehydrated and stained. Replace ultrapure water in **steps 2** and **4** every three slides and fresh 95% ethanol and 100% ethanol every two slides. All reagents should be kept RNase free.

5. After all slides dehydrated and stained, in a chemical hood, quickly and carefully blot edge of slides on Kimwipes to remove excess xylene.
6. Transfer the slides to a slide box containing desiccators and proceed with laser capture microdissection (*see* **Note 8**).

3.3.2 Frozen Section Cresyl Violet Staining

Cresyl violet is a basic dye used to stain Nissl substance in the cytoplasm of neurons purple-blue. It can also stain the granules in hyaline cartilage, thus useful for identifying the growth plate [18]. Cartilages are stained purple-blue by cresyl violet while mineralized tissues are stained gray or black. Cresyl violet staining can be used as an alternative to H&E Staining (*see* **Note 9**) (Fig. 2).

1. Prepare 50 ml volumes of the following solutions in six separate 50 ml conical tubes:

(a) 0.1% cresyl violet in 50% ethanol.

(b) 70% ethanol (2 tubes).

(c) 95% ethanol.

(d) 100% ethanol.

(e) Ultrapure water (1 tube).

2. Prepare 250 ml xylene in a Tissue-Tek Slide Staining Dish (solvent-resistant green dish) and Staining Rack in a chemical hood.

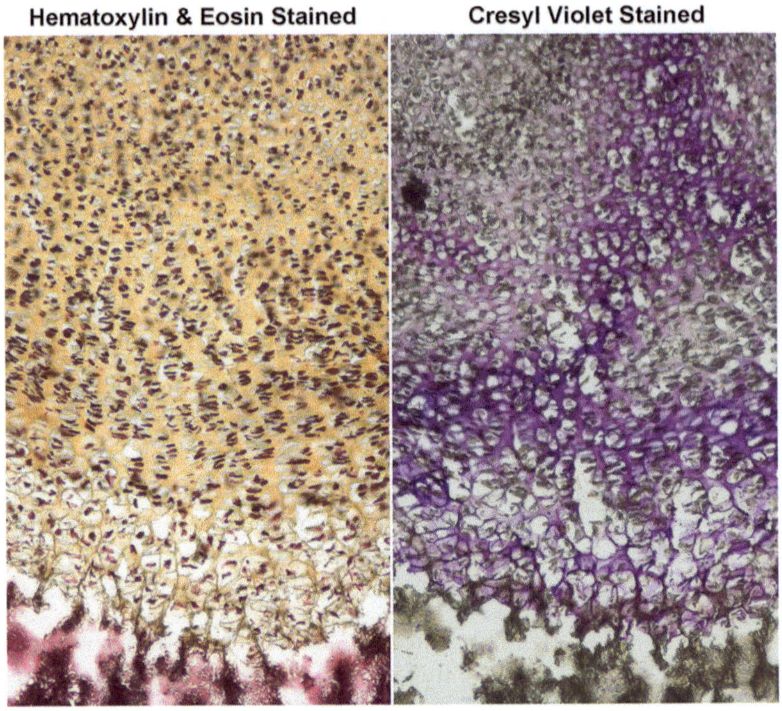

Fig. 2 Staining of 1-week-old mouse proximal tibial growth plate with hematoxylin and eosin (left) or cresyl violet (right)

3. Cool a slide box on dry ice. Immediately before staining, obtain frozen sectioned slides from −70 °C and transfer to the cooled slide box on dry ice.
4. Use forceps to dehydrate and stain slides individually, as follows:
   (a) 70% ethanol, 30 s.
   (b) ultrapure water, 30 s, use forceps to move slide up and down to remove OCT.
   (c) 0.1% cresyl violet in 50% ethanol, 30 s.
   (d) 70% ethanol, 15 s.
   (e) 95% ethanol, 30 s.
   (f) 100% ethanol 30 s.
   (g) Xylene, 5 min or hold.

   Perform the last step in a chemical hood, keep slides in xylene and start staining the next slide. Keep all slides in xylene until all slides are dehydrated and stained. Replace ultrapure water in **step 2** every three slides and fresh 95% ethanol and 100% ethanol every two slides. All reagents should be kept RNase free.

5. After all slides dehydrated and stained, in a chemical hood, quickly and carefully blot edge of slides on Kimwipes to remove excess xylene.
6. Transfer the slides to a slide box containing desiccators and proceed with laser capture microdissection.

3.4 Laser Capture Microdissection

The Arcturus^{XT} system uses an IR laser beam and UV cutting laser to procure desired cells regions for further analysis. This system allows the user to locate particular regions of sections, control LCM cap placement on the tissue, precisely delineate a cell population for procurement, and finally capture and cut out the selected cells.

3.4.1 Preparation of Lysis Buffer Solution

1. Prepare appropriate volume of lysis buffer as follows: dilute 1 part BME in 9 parts of RLT buffer. For each cap, 50 ul of buffer is required (5 μl BME + 45 μl RLT buffer) (*see* **Note 10**).

3.4.2 Computer Set-Up and Software Settings

1. Power on the microscope.
2. Turn on the computer.
3. Power on the Arcturus^{XT} instrument.
4. Open the Arcturus^{XT} software on computer.
5. Configure settings in program by navigating to the tool panel on the right side of the screen. Under the **inspect** tab, change diffuser setting to **in**. This allows the white light to be dispersed evenly across the image.
6. Under the **select** tab, select the i to create a custom circle. The size of the CapSure LCM cap is 2900 microns.
7. The bottom of the program, from left to right, contains the cap and slide handling areas, slide overview image, and the QC caps indicator. Choose the appropriate cap settings: HS, FRM. Mark how many caps you will place in the machine.

3.4.3 Loading Caps and Slide Samples

1. Load CapSure LCM caps on the stage.
2. Obtain first slide to dissect and sandwich it adjacent to a 75 × 25 mm Superfrost Plus slide. Place this sandwich in any slide holder of the dissecting scope. *Note*: dissecting one slide at a time prevents tissue rehydration.
3. On the bottom left of the program, click on the corresponding letter next to the slide holder containing your slide to move the stage inward.
4. Right-click in the gray box at the bottom of the program and choose **remember settings**.
5. Right-click again in the gray box at the bottom of the program and select **reacquire overview image** to obtain a screenshot of

the slide in the program. This will help navigate between different sections on the same slide.

6. Under the **inspect** tab, change brightness and focus image if necessary.

3.4.4 Cap Placement and Marking Region of Interest

1. Select any capture group (A, B, C, or D) under the **select** tab. Place the custom circle from one capture group at 2× magnification. Center the cap by clicking on the number associated capture group.
2. Switch to 10× magnification and place the cap by selecting on the cap placement icon.
3. Locate IR spot.
4. Select a different capture group. Using the freehand draw feature, trace a closed section of interest around the targeted population of homogeneous cells. IRspots should automatically appear in the marked section if a closed section was drawn.
5. The resting zone is distinguished by small round chondrocytes that appear between the secondary ossification center and proliferating columnar cells.
6. The proliferative zone is identified by elongated cells that resemble columns arranged in a parallel fashion.
7. The beginning of cell hypertrophy, which represented the prehypertrophic zone, is avoided.
8. The region from the end of prehypertrophic cells to the region right above the terminal hypertrophic cells is characterized as the hypertrophic zone (Fig. 3).

3.4.5 Capture and Cut

1. Select the capture and cut icon under the **microdissect** tab.
2. If the cut is not uniform or does not penetrate the tissue well, recut the area by clicking on the cut only icon (*see* **Note 11**).
3. Right-click on the numbers in the capture group and delete them, if you made multiple sections on a single slide and will repeat cut on other sections.
4. Once all sections are dissected and adhered to the cap, click **QC** icon under the **microdissect** tab to move the cap to the QC station.
5. Select present stage under the **setup** tab to move the stage out and inspect the cap for the microdissected tissue.
6. Add 50 μl of lysis buffer (1:10 beta-mercaptoethanol in RLT buffer) to a 0.5 ml domed cap, Gene-Amp Thin-Walled Reaction tube.
7. Firmly attach the LCM cap to the open tube. Invert the tube and flick the liquid contents to cover the tissue on the LCM cap.

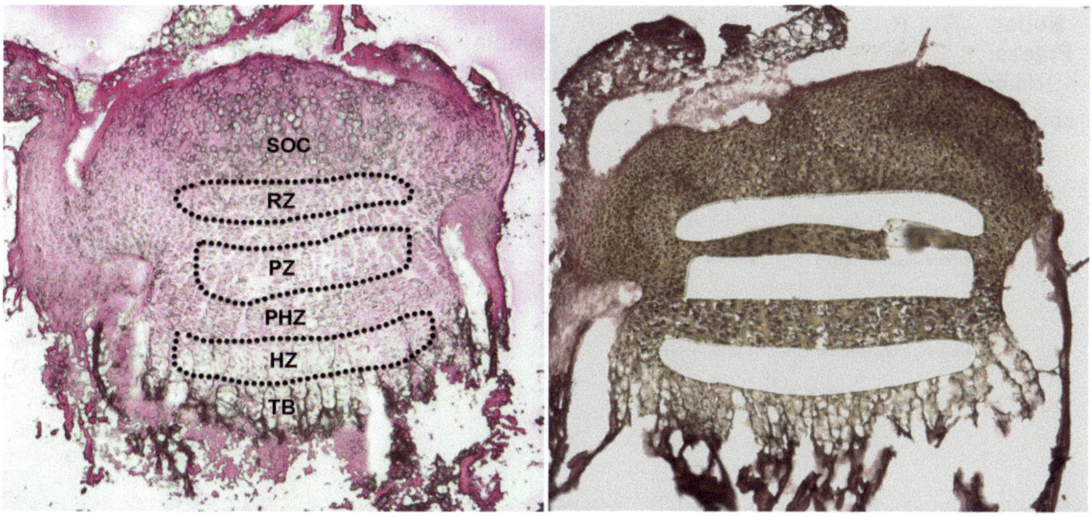

Fig. 3 Growth plate dissection by LCM. Left: Eosin-stained frozen section of 1-week-old mouse proximal tibial cartilage showing the location of secondary ossification center (SOC), resting zone (RZ), proliferative zone (PZ), prehypertrophic zone (PHZ), hypertrophic zone (HZ), and trabecular bone (TB). RZ, PZ, and HZ were circled with dotted line indicating areas to be dissected with LCM. Right: Hematoxylin & eosin–stained frozen section of 1-week-old mouse proximal tibial cartilage after LCM

8. Place on dry ice until all prestained slides are completed (*see* **Note 12**).
9. Store tubes at −70 °C until ready for RNA extraction.

3.5 RNA Extraction

1. Obtain samples from −70 °C and place them in a tube rack.
2. Invert the tube rack and allow samples to thaw onto LCM cap for 30 min at room temperature.
3. Spin down the tubes and discard LCM cap.
4. Use micro RNA extraction kit to extract RNA from samples, combining replicates of each biological specimen.
5. RNA concentration and integrity can be assessed using the Agilent 2100 Bioanalyzer with an RNA Pico Chips.
6. RNA can be used for downstream applications such as reverse transcription and RNA sequencing.

4 Notes

1. Tissue detachment is a commonly encountered issue during histological staining. Since most animal tissues carry an excess of acidic amino acids, precoating the PEN membrane frame slides with a basic polymer such as poly-L-lysine maintains tissue adhesion to the polyanionic cell surfaces [19].

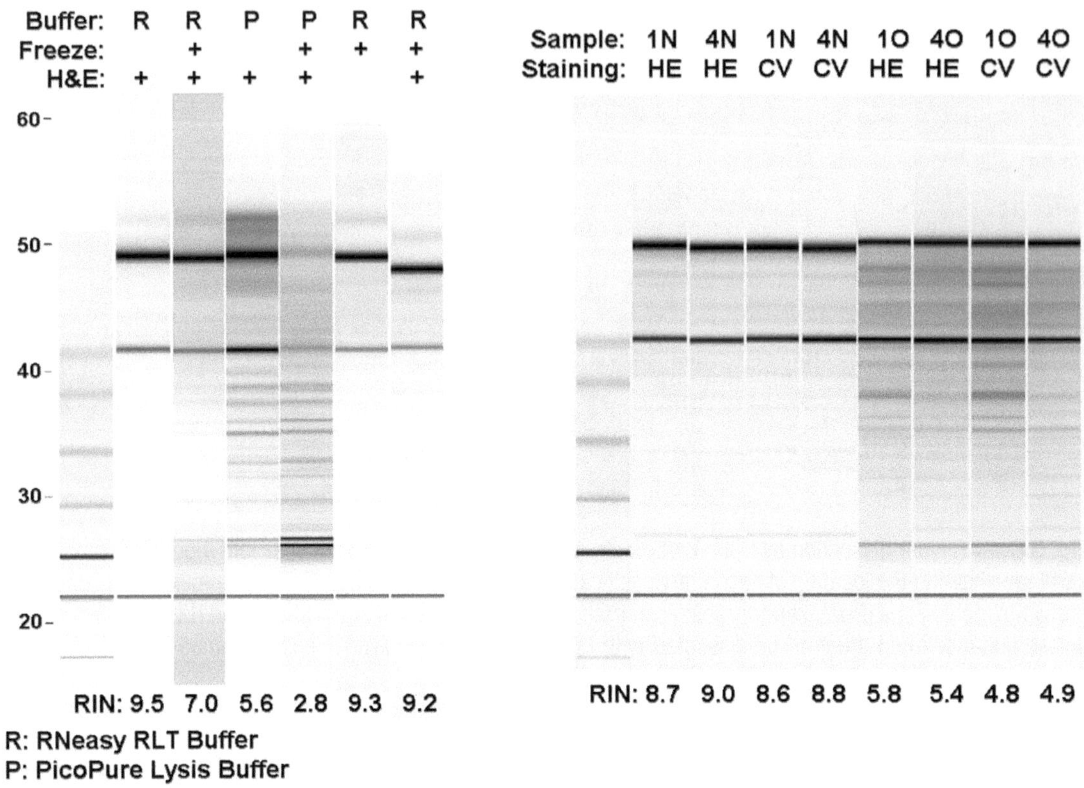

Fig. 4 RNA quality of LCM samples under different conditions. *Left panel*: we compared the quality of RNA isolated with RNeasy micro kit (R) and PicoPure RNA isolation kit (P) (lanes 1–4) and found significantly higher RNA quality from the RNeasy micro kit. Freezing the LCM sample immediately after dissection in the PicoPure lysis buffer led to almost complete RNA degradation (lane 4); however, negligible RNA degradation was observed with the RNeasy micro kit. We also found that H&E staining (lane 5) did not negatively affect RNA quality in comparison to no staining (lane 6). *Right panel*: we compared the RNA quality isolated from: different ages of mice (1 week versus 4 weeks old); freshly dissected samples (new (N)) versus samples stored in OCT compound at −70 °C for over a year (old (O)); and hematoxylin and eosin (HE) stained or cresyl violet (CV) stained sections. We found that the age of the animal and staining method did not significantly affect RNA quality, but found noticeable degradation after long term storage of samples at −70 °C

2. For sectioning of undecalcified bones from older animals, we recommend using the Kawamoto's Film Method [20].
3. Frozen tissue block can be stored at −70 °C short term with minimal RNA degradation. However, we found noticeable decrease in RNA integrity (RNA integrity number (RIN) decreased to 5.0–7.0) if the samples were stored for more than a year (*see* Fig. 4).
4. The PEN membrane slide is basically a piece of membrane fixed on a metal frame. On its own, the membrane does not provide the temperature difference necessary for the sectioned tissue to thaw and stick to the membrane. Therefore, a frame supporting

slide must be placed below the PEN membrane slide to warm the membrane that is in contact with the sectioned tissue. A frame supporting slide is a slide with an elevated area (1 mm height × 16 mm width × 44 mm length) that fits into the recessed area of the PEN membrane slide. It could be custom made by trimming a regular glass slide into 16 mm by 44 mm and adhering it by superglue to another regular glass slide. Alternatively, it could also be produced by 3D printing using a suitable material.

5. Placing multiple sections on a single PEN membrane slides can reduce both the number of slides as well as number of caps used in LCM for each sample, thus saving time and cost per sample.
6. Unlike the frozen tissue block, which is relatively stable at −70 °C, RNA in sectioned tissue were not as stable and thus should be used for LCM within a couple of days.
7. Washing off OCT compound in ultrapure water is a critical step in staining prior to LCM. The presence of OCT compound on the slide or on the tissue can interfere with microdissection and RNA isolation.
8. Ideally, each slide should be stained and immediately proceed to laser capture microdissection. Alternative, several slides can be stained, dried and kept in a slide box containing desiccators, while performing LCM on the stained slides one at a time, yields comparable RNA integrity and concentration.
9. Cresyl violet staining procedure involve fewer aqueous steps, which in theory, lower the chance of RNA degradation during staining. However, the RNA integrity obtained using either staining methods appears to be comparable. We also found that H&E staining itself does not negatively impact RNA integrity (*see* Fig. 4).
10. Isolating RNA from microdissected samples involves dealing with low cell numbers. We have extensively tested different RNA extraction kits designed for LCM. We found that the PicoPure RNA Isolation Kit, which is originally designed and produced by Arcturus (the same company producing the LCM machine), consistently lead to partial RNA degradation. We found that the degradation occurred during a 30 min 42 °C on-cap digestion step at the beginning of the procedure. We have subsequently switched to the QIAGEN RNeasy Micro Kit, which yielded higher RNA quality and concentration.
11. Repeated recutting using the UV laser could lead to RNA degradation due to overheating in adjacent tissues and thus should be used with caution.

12. We found that freezing and storing the dissected tissue in RLT buffer has minimal impact on RNA integrity. In contrast, we found that immediately freezing the dissected tissue in lysis buffer from the PicoPure RNA isolation kit is detrimental to RNA quality (*see* Fig. 4).

References

1. Kronenberg HM (2003) Developmental regulation of the growth plate. Nature 423 (6937):332–336. https://doi.org/10.1038/nature01657
2. Lui JC, Nilsson O, Baron J (2014) Recent research on the growth plate: recent insights into the regulation of the growth plate. J Mol Endocrinol 53(1):T1–T9. https://doi.org/10.1530/jme-14-0022
3. Mizuhashi K, Ono W, Matsushita Y, Sakagami N, Takahashi A, Saunders TL, Nagasawa T, Kronenberg HM, Ono N (2018) Resting zone of the growth plate houses a unique class of skeletal stem cells. Nature 563(7730):254–258. https://doi.org/10.1038/s41586-018-0662-5
4. Abad V, Meyers JL, Weise M, Gafni RI, Barnes KM, Nilsson O, Bacher JD, Baron J (2002) The role of the resting zone in growth plate chondrogenesis. Endocrinology 143 (5):1851–1857. https://doi.org/10.1210/endo.143.5.8776
5. Ono N, Ono W, Nagasawa T, Kronenberg HM (2014) A subset of chondrogenic cells provides early mesenchymal progenitors in growing bones. Nat Cell Biol 16(12):1157–1167. https://doi.org/10.1038/ncb3067
6. Lui JC, Yue S, Lee A, Kikani B, Temnycky A, Barnes KM, Baron J (2019) Persistent Sox9 expression in hypertrophic chondrocytes suppresses transdifferentiation into osteoblasts. Bone 125:169–177. https://doi.org/10.1016/j.bone.2019.05.027
7. Dy P, Wang W, Bhattaram P, Wang Q, Wang L, Ballock RT, Lefebvre V (2012) Sox9 directs hypertrophic maturation and blocks osteoblast differentiation of growth plate chondrocytes. Dev Cell 22(3):597–609. https://doi.org/10.1016/j.devcel.2011.12.024
8. Garrison P, Yue S, Hanson J, Baron J, Lui JC (2017) Spatial regulation of bone morphogenetic proteins (BMPs) in postnatal articular and growth plate cartilage. PLoS One 12(5): e0176752. https://doi.org/10.1371/journal.pone.0176752
9. Li Y, Dudley AT (2009) Noncanonical frizzled signaling regulates cell polarity of growth plate chondrocytes. Development 136 (7):1083–1092. https://doi.org/10.1242/dev.023820
10. Emmert-Buck MR, Bonner RF, Smith PD, Chuaqui RF, Zhuang Z, Goldstein SR, Weiss RA, Liotta LA (1996) Laser capture microdissection. Science 274(5289):998–1001. https://doi.org/10.1126/science.274.5289.998
11. Mikulowska-Mennis A, Taylor TB, Vishnu P, Michie SA, Raja R, Horner N, Kunitake ST (2002) High-quality RNA from cells isolated by laser capture microdissection. BioTechniques 33(1):176–179. https://doi.org/10.2144/02331md06
12. Datta S, Malhotra L, Dickerson R, Chaffee S, Sen CK, Roy S (2015) Laser capture microdissection: big data from small samples. Histol Histopathol 30(11):1255–1269. https://doi.org/10.14670/hh-11-622
13. Espina V, Wulfkuhle JD, Calvert VS, VanMeter A, Zhou W, Coukos G, Geho DH, Petricoin EF 3rd, Liotta LA (2006) Laser-capture microdissection. Nat Protoc 1 (2):586–603. https://doi.org/10.1038/nprot.2006.85
14. Nichterwitz S, Chen G, Aguila Benitez J, Yilmaz M, Storvall H, Cao M, Sandberg R, Deng Q, Hedlund E (2016) Laser capture microscopy coupled with smart-seq2 for precise spatial transcriptomic profiling. Nat Commun 7:12139. https://doi.org/10.1038/ncomms12139
15. Curran S, McKay JA, McLeod HL, Murray GI (2000) Laser capture microscopy. Mol Pathol 53(2):64–68. https://doi.org/10.1136/mp.53.2.64
16. Lui JC, Garrison P, Nguyen Q, Ad M, Keembiyehetty C, Chen W, Jee YH, Landman E, Nilsson O, Barnes KM, Baron J (2016) EZH1 and EZH2 promote skeletal growth by repressing inhibitors of chondrocyte proliferation and hypertrophy. Nat Commun

7:13685. https://doi.org/10.1038/ncomms13685

17. Lui JC, Jee YH, Garrison P, Iben JR, Yue S, Ad M, Nguyen Q, Kikani B, Wakabayashi Y, Baron J (2018) Differential aging of growth plate cartilage underlies differences in bone length and thus helps determine skeletal proportions. PLoS Biol 16(7):e2005263. https://doi.org/10.1371/journal.pbio.2005263
18. Takahashi N, Tarumi W, Hamada N, Ishizuka B, Itoh MT (2017) Cresyl violet stains mast cells selectively: its application to counterstaining in immunohistochemistry. Zool Sci 34 (2):147–150. https://doi.org/10.2108/zs160162
19. Mazia D, Schatten G, Sale W (1975) Adhesion of cells to surfaces coated with polylysine. Applications to electron microscopy. J Cell Biol 66(1):198–200. https://doi.org/10.1083/jcb.66.1.198
20. Kawamoto T, Kawamoto K (2014) Preparation of thin frozen sections from nonfixed and undecalcified hard tissues using Kawamot's film method (2012). Methods Mol Biol 1130:149–164. https://doi.org/10.1007/978-1-62703-989-5_11

Chapter 9

RNA Isolation from Articular Cartilage Tissue

Hongjun Zheng and Audrey McAlinden

Abstract

Isolation of high-quality RNA directly from tissues is desirable to obtain precise information of in vivo gene expression profiles in cells embedded within their extracellular matrix (ECM). It is well known that purification of RNA from cartilage tissues is particularly challenging due to low cell (chondrocyte) content and its dense ECM rich in large negatively charged proteoglycans that can copurify with RNA. Older methodologies to purify RNA from cartilage involved the use of concentrated denaturing solutions containing guanidinium isothiocyanate followed by ultracentrifugation in cesium trifluoroacetate. Such ultracentrifugation approaches are rarely used now since the emergence of more user-friendly mini spin column chromatography kits. For this chapter, we tested and compared three methods to isolate RNA from immature murine articular (femoral head) cartilage and found that the combination of TRIzol® reagent and spin column chromatography (Norgen Total RNA Purification Kit) was the best approach to generate higher quality RNA. Here, the average RNA Integrity Number (RIN), as determined by Bioanalyzer technology, was 7.1. We then applied this method to attempt to isolate RNA directly from human articular cartilage harvested from three osteoarthritic (OA) knee joint specimens. As expected, the concentration and quality of RNA obtained differed between samples. However, from one specimen, we were able to isolate approximately 3 μg of total RNA (including small noncoding RNAs) from 100 mg of human OA cartilage with a RIN = 7.9. Despite the patient-to-patient variabilities that are known to exist between cartilage specimens from OA joints, we have demonstrated that it is possible to obtain reasonably high levels of RNA from human OA articular cartilage at a quality suitable for downstream analyses including microarray and RNA-Seq. A detailed description of our preferred RNA purification methodology, which can be used to isolate RNA from human, bovine, or rodent cartilage tissue, is provided in this chapter.

Key words Cartilage, Articular cartilage, RNA, RNA purification, Tissue pulverization, TRIzol®, Norgen kit, Spin column, Bioanalyzer, NanoDrop™, Chondrocyte, Osteoarthritis

1 Introduction

Isolation of high-quality RNA is critical for a number of downstream approaches to determine gene expression levels within cells of interest. Such approaches include cDNA synthesis, RT-qPCR, digital PCR, microarray analysis, and RNA-Seq. A number of relatively straightforward techniques to isolate RNA from primary cells or cell lines exist, including organic extraction methods involving phenol-containing solutions, solid phase silica-based spin column

Tariq M. Haqqi and Véronique Lefebvre (eds.), *Chondrocytes: Methods and Protocols*, Methods in Molecular Biology, vol. 2245, https://doi.org/10.1007/978-1-0716-1119-7_9,

approaches, or techniques utilizing magnetic beads. All of these approaches involve addition of a lytic reagent or denaturant that should be in contact with the cellular contents at the moment when the cells are disrupted. This can be more challenging to achieve in cases where the goal is to isolate RNA directly from tissues where cells are embedded within a complex extracellular matrix (ECM). RNA isolation can be even more problematic in scenarios where tissues are relatively "hard" such as bone and cartilage. Bone tissue is very rigid due to the high mineral content in the ECM. Articular (hyaline) cartilage, that lines the ends of bones at various joints in the body, is not as rigid as bone yet it is a tough, fairly flexible rubber-like tissue that facilitates the transmission of loads within the joints. The critical weight-bearing properties of articular cartilage are due to the unique composition of its ECM that is composed primarily of type II collagen fibers and proteoglycans (PGs) that function to resist tensile and compressive loads, respectively [1–3]. In addition to the low cell content of mature articular cartilage, additional challenges in isolating high-quality RNA from such tissue are due to its highly cross-linked ECM, particularly the presence of abundant, negatively charged PGs that can copurify with RNA. Furthermore, cartilage must undergo a harsh pulverization process which can also compromise RNA integrity.

Obtaining RNA directly from cartilage tissue at a concentration and quality suitable for downstream analysis is appealing since it provides a better representation of gene expression profiles in vivo. This may be important in scenarios where the goal is to characterize potential changes in cartilage tissue gene expression from conditional knock-out or transgenic mice, for example. Or if gene expression signatures are to be compared in human cartilage from different joints, or from patients with specific conditions affecting cartilage (e.g., osteoarthritis, type II diabetes), then analysis of RNA obtained from the tissue would be preferable. While enzymatic-based methods are well-established for isolating primary chondrocytes from cartilage tissue, such procedures and downstream culturing techniques will undoubtedly induce cellular responses to alter transcriptional processes and, hence, gene expression patterns. In fact, alterations in mRNA as well as small microRNA (miRNA) expression patterns have been demonstrated when comparing RNA isolated from cartilage tissue versus extracted chondrocytes [4, 5].

Our laboratory has a lot of experience in working with murine, bovine and human cartilage and we have applied various methods over the years to attempt to isolate RNA from these tissues. In this chapter, we report on the method that works best in our hands with respect to reproducibility as well as quantity and quality of RNA obtained. This method involves pulverization of frozen cartilage tissue followed by organic extraction and spin column chromatography. Specifically, we use TRIzol® reagent (a mixture of guanidine isothiocyanate, phenol, and other proprietary components) followed by two chloroform extraction steps and the Norgen Total RNA Purification Kit (Norgen Biotek Corp., Canada). Purification using the Norgen Kit is based on mini spin column chromatography using the manufacturer's proprietary resin as the separation matrix. Older methods have utilized a guanidinium isothiocyanate buffer (similar to TRIzol®) followed by cesium trifluoroacetate ultracentrifugation as a means to isolate RNA from its dense PG-rich ECM [6–8]. In these cases, more straightforward mini spin column methods were not available or were not thoroughly tested for their usefulness in the isolation of RNA from cartilage. Also, Bioanalyzer (an automated chip-based capillary electrophoresis machine) technology was not available to precisely assess the quality of RNA isolated from cartilage using these older protocols. For the purpose of this chapter, we compared three methods to isolate RNA from immature (2–3 weeks) murine femoral head cartilage: (1) TRIzol®/Norgen Kit, (2) Phenol-SDS/Norgen Kit, and (3) Norgen Kit/Proteinase K. Methods (1) and (2) are almost identical apart from the use of a phenol–SDS solution instead of TRIzol®. Utilization of a phenol–SDS-containing buffer has been reported for extraction of RNA from plant material containing high levels of polysaccharides [9]. Therefore, we aimed to test if RNA extraction from cartilage could be improved with this buffer given the high levels of PGs in this tissue. In Method (3), we omitted the organic extraction steps and utilized the Norgen Kit only, with the addition of Proteinase K to the kit's lysis buffer.

Table 1 shows the concentration, purity, and integrity of RNA isolated from immature murine femoral head cartilage using the three different methods as indicated. In each sample, femoral head cartilage was pooled from 4 mice resulting in approximately 25 mg of tissue. Spectrophotometry (NanoDrop™) technology provided information on RNA quantity as well as purity (i.e., A_{260}/A_{280} and $A_{260/230}$ values). A_{260}/A_{280} values ~2 generally indicates "pure" RNA. Similarly, an A_{260}/A_{230} value close to 2 generally indicates low to no contamination. High A_{230} values in relation to A_{260} are commonly a result of TRIzol® or lysis buffer component contamination. For cartilage samples, low A_{260}/A_{230} values may also be due to some proteoglycan contamination, but we predict that most of the PGs have been denatured and removed by the TRIzol® and chloroform extraction steps as well as the spin column chromatography steps in our approach. Table 1 shows that the A_{260}/A_{280} values for each sample were all approximately 2, regardless of the method used. However, there was very high variability in A_{260}/A_{230} values between samples. The lowest A_{260}/A_{230} values were found in the RNA samples obtained using the Norgen kit/Proteinase K method and this corresponded to the lowest RIN values as well. RNA samples with RINs of 7 or higher are generally considered to be of good quality (i.e., mostly intact RNA) and suitable for

Table 1
Concentration, purity, and integrity of RNA isolated from immature murine femoral head cartilage using different methods

Method	Murine Femoral Head Cartilage	Total RNA (μg/25 mg Tissue)	A_{260}/A_{280}	A_{260}/A_{230}	RIN
TRIzol® + Norgen	Sample 1	9.27	2.13	0.86	6.6
	Sample 2	9.06	2.15	2.19	7.4
	Sample 3	8.79	2.15	1.34	7.2
SDS/Phenol + Norgen	Sample 4	8.75	2.03	1.3	3.1
	Sample 5	5.92	2.05	2.09	5.6
	Sample 6	10.63	2.04	1.91	3.6
Norgen + Proteinase K	Sample 7	6.76	1.95	0.37	3.0
	Sample 8	4.84	2.02	0.6	2.5
	Sample 9	6.02	2.04	0.67	2.4

In each sample, hip cartilage from 4 mice at 2–3 weeks of age was pooled (i.e., cartilage from 8 femoral heads) resulting in approximately 25 mg of tissue. Table shows NanoDrop™ spectrophotometry results which provide information on the purity (i.e., A_{260}/A_{280} and A_{260}/A_{230} ratios) as well as the total amount of RNA extracted from each sample. The RIN value for each sample, as determined by a Bioanalyzer, is also shown. *See* Fig. 1 for corresponding Bioanalyzer-derived gels and plots for each sample

downstream analyses including real time qPCR, microarray and RNA-Seq. The best RIN values were obtained using the TRIzol®/Norgen kit method (Table 1). Figure 1 shows the corresponding Bioanalyzer-derived electrophoretic profiles of each sample listed in Table 1. Samples isolated with the TRIzol®/Norgen kit method showed clear peaks/bands corresponding to 28S and 18S ribosomal RNA and low levels of smaller molecular weight bands indicating RNA degradation or contamination (Fig. 1a–c). Electrophoretic profiles of RNA isolated using the other two methods clearly showed higher levels of degraded RNA and/or proteoglycan contamination (Fig. 1d–i), especially in the samples isolated using the Norgen Kit/Proteinase K method (Fig. 1g–i). We predict that this latter method resulted in issues with high levels of PG contamination due to omitting the organic extraction steps. There are other reports describing a spin column chromatography method (Qiagen RNeasy Kit) to isolate RNA from human articular cartilage without prior TRIzol® and chloroform extraction steps [10, 11]. However, Bioanalyzer analysis of resulting RNA was not carried out or available at that time. In conclusion, the TRIzol®/Norgen Kit method was the best approach to isolate abundant RNA (~9 μg/25 mg murine cartilage tissue) that was mostly intact as indicated by the RIN values. Note that immature, developing cartilage samples (whether murine, bovine, or human) have a higher cell content than fully developed cartilage tissues. Therefore, one should expect an overall lower RNA yield from more hypocellular mature articular cartilage.

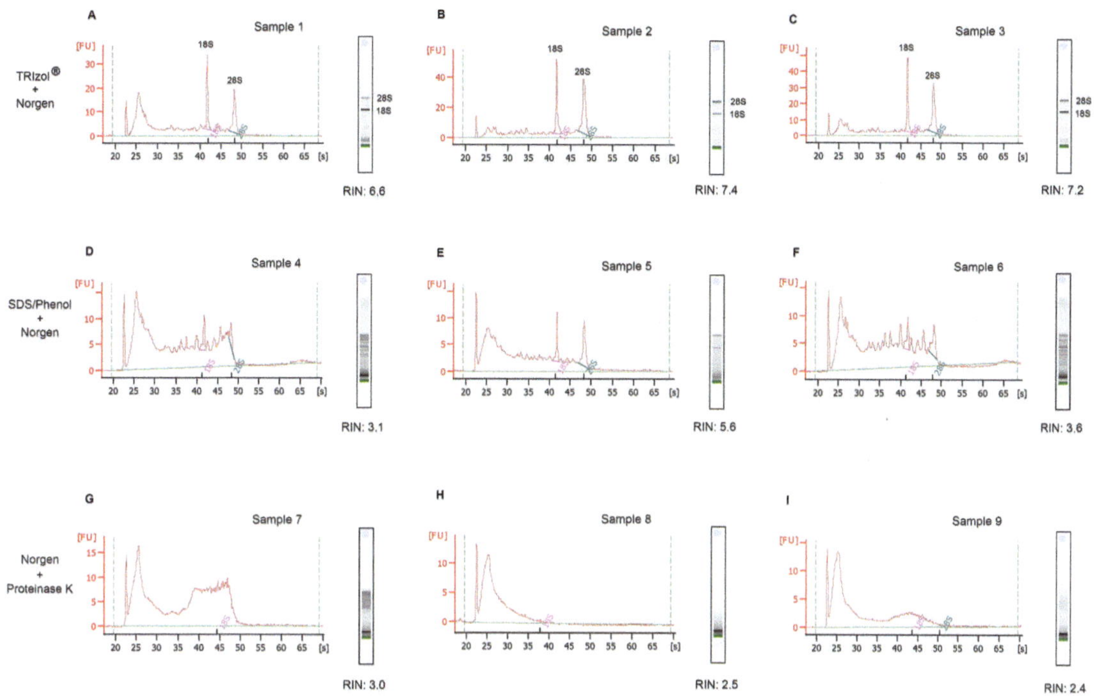

Fig. 1 Electrophoretic profiles of RNA extracted from murine immature femoral head cartilage. In each sample, hip cartilage from 4 mice at 2–3 weeks of age was pooled (i.e., cartilage from 8 femoral heads) resulting in approximately 25 mg of tissue. RNA integrity data, including the RIN values, are shown for samples where RNA was isolated using TRIzol®/Norgen kit (**a–c**), SDS-Phenol/Norgen kit (**d–f**) or Norgen kit/Proteinase K (**g–i**). *See* Table 1 for corresponding NanoDrop™ spectrophotometer data for each sample

Clearly, the A_{260}/A_{280} values can be misleading since all samples listed in Table 1 showed ratios close to 2, but many of these samples had very low RINs This highlights the importance of analyzing RNA by both spectrophotometry and Bioanalyzer technology.

We then used the TRIzol®/Norgen Kit to isolate RNA from three human knee OA articular cartilage specimens. Table 2 shows similar A_{260}/A_{280} values of ~2 for all samples, but there were differences in RNA quantity, A_{260}/A_{230} ratios and RINs (Fig. 2a–c) between the samples. For "OA Sample 3," ~3 μg of RNA was obtained from 100 mg of tissue with a RIN = 7.9. This is a very impressive result for RNA obtained from human OA cartilage. Although the A_{260}/A_{230} was low, we were somewhat less concerned about this (*see* **Note 1**). The variability in results obtained with the other two specimens with respect to total RNA isolated and RIN values is not altogether surprising given the patient-to-patient variability in cartilage specimens (e.g., differing degrees of cartilage degradation) received from the operating room.

We also carried out qPCR to determine relative expression levels of two common chondrocyte marker genes (type II collagen

Table 2
Concentration, purity, and integrity of RNA isolated from human osteoarthritic articular cartilage using a combination of TRIzol® reagent and the Total RNA Purification Kit (Norgen)

Method	Human Osteoarthritic Cartilage	Total RNA (μg/100 mg Tissue)	A_{260}/A_{280}	A_{260}/A_{230}	RIN
TRIzol® + Norgen	OA sample 1	1.37	2.09	1.63	5.5
	OA sample 2	0.99	2.11	1.01	4.3
	OA sample 3	2.91	2.16	0.88	7.9

From three independent OA knee joint specimens, 100 mg of tissue was harvested. Table shows NanoDrop™ spectrophotometer data which provides information on RNA purity (A_{260}/A_{280} and A_{260}/A_{280} ratios) and concentration (i.e., total amount of RNA isolated from 100 mg tissue). The RIN values, as determined by a Bioanalyzer, are also shown. *See* Fig. 2 for corresponding Bioanalyzer-derived gels and plots for each sample

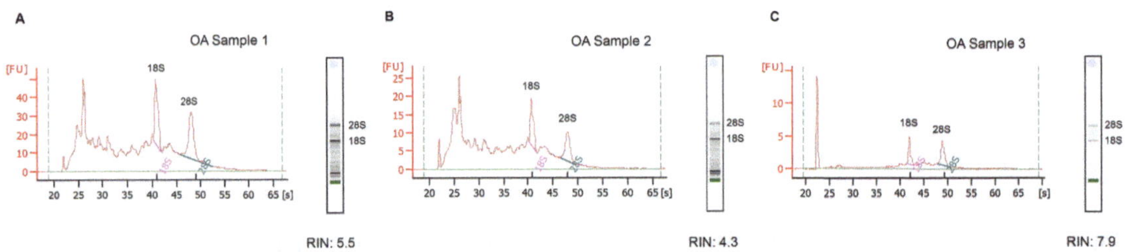

Fig. 2 Electrophoretic profiles of RNA extracted from human osteoarthritic articular cartilage using a combination of TRIzol® reagent and the Total RNA Purification Kit (Norgen). From three independent OA knee joint specimens, 100 mg of tissue was harvested. RNA integrity data, including the RIN values, are shown for each of the three samples (**a–c**). *See* Table 2 for corresponding NanoDrop™ spectrophotometer data for each sample

and microRNA-140) in the murine and human RNA samples obtained by the TRIzol®/Norgen kit method. High expression of *COL2A1* was detected, relative to *GAPDH*, as expected (Fig. 3a, b). We also detected the presence of miR-140, one of the best-described and most highly expressed non-coding miRNAs in chondrocytes (Fig. 3c, d) [12]. This data demonstrates that the methodology we applied was also successful in isolating small non-coding RNAs as well as mRNAs and long non-coding RNAs.

With respect to human OA cartilage, two other studies have reported success in isolating good quality RNA. Ruettger et al. utilized a phenol–chloroform-free filter–based method (RNA*queous*™ kit) and obtained a RIN value of 6.2 [13]. However, it should be noted that small RNAs were not isolated using this method. While this protocol can be modified to isolate small RNAs by the addition of more ethanol to the lysis buffer, this may likely affect RNA yield and potentially increase proteoglycan contamination. The recent study by Le Bleu et al. [14] utilized the traditional TRIzol® method with a modified step involving addition of high salt to the aqueous layer prior to RNA precipitation.

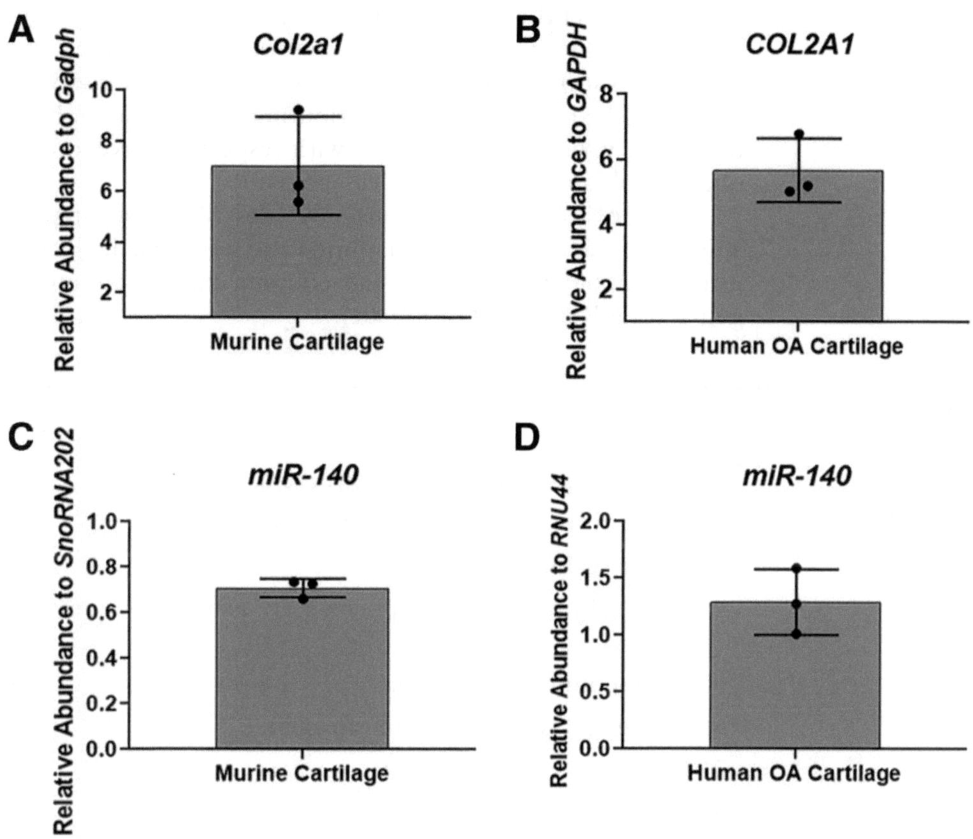

Fig. 3 Expression levels of type II collagen mRNA and microRNA-140 in total RNA isolated from murine and human OA articular cartilage using TRIzol® and the Norgen Total RNA Purification Kit. Relative expression levels of type II collagen mRNA (*Col2a1*) and miR-140 in cartilage from immature murine femoral heads (**a** and **c**). Expression of *COL2A1* and miR-140 in human OA cartilage (**b**, **d**). Relative expression levels are shown as a mean of three independent specimens

This is a modification recommended in the manufacturer's protocol for samples containing high levels of proteoglycans. Impressive RIN values (7.9 and 8.4) were obtained and RNA yield from 100 mg of tissue was similar to our findings. A_{260}/A_{230} ratios were also impressive. It is however, unclear if small RNAs were also isolated using this methodology.

Another important factor which most likely resulted in their ability to isolate mostly intact, pure RNA from OA cartilage is due to the fact that they were able to obtain tissue and freeze it within minutes following surgery in the operating room. Many laboratories, including ours, do not have such a system in place for obtaining human tissue immediately after removal from the patient, whether from diseased or non-diseased joints. We usually obtain cartilage tissue within 1–3 h following surgery in sterile containers without any media or PBS. Undoubtedly, the period of time cartilage tissue is stored ex vivo prior to transportation to the laboratory will likely affect the chondrocyte response and hence RNA quality/yield. In many cases, we cannot control how long each cartilage specimen is stored in the operating room following surgery and this is likely to be a major factor contributing toward the high patient-to-patient variability we observe with respect to RNA quality and quantity obtained from OA cartilage tissue.

In the following sections, we describe our step-by-step approach to isolate RNA from murine and human articular cartilage using TRIzol® reagent and spin column chromatography (Norgen's Total RNA Purification kit).

2 Materials

2.1 Tools and Equipment

1. Sterile surgical scissors and tweezers.
2. Sterile disposable scalpel blades (size #10).
3. Sterile scalpel blade holders.
4. Personal protective equipment (lab coat, sterile gloves, safety glasses).
5. Biopsy punch (4–6 mm).
6. Cryogenic vials and liquid nitrogen.
7. Freezer (−80 °C).
8. Fume hood.
9. Mikro-dismembrator, Model U (B. Braun Biotech International; Sartorius).
10. Mikro-dismembrator accessories: (1) shaking flasks (7 ml stainless steel), (2) grinding balls (10 mm chromium steel) (Sartorius).
11. Weighing scale (accuracy = 0.1 mg).
12. Ice bucket or suitable container to hold liquid nitrogen.
13. Refrigerated high-speed tabletop microcentrifuge.
14. Microcentrifuge tubes (1.5 ml and 2 ml; RNase free).
15. Pipette tips (10 μl, 200 μl, 1 ml; RNase free).
16. Spectrophotometer (NanoDrop™).
17. Bioanalyzer (Agilent).

2.2 Reagents

1. RNaseZap® or RNase AWAY®.
2. Sterile PBS.
3. TRIzol® reagent.
4. Chloroform.
5. Ethanol (200 proof).
6. Isopropanol.

7. Norgen Total RNA Purification Kit (Cat. #37500; Norgen Biotek Corp., Canada).
8. RNAse-Free DNase I kit (Norgen Biotek Corp., Canada).

3 Methods

Prior to tissue harvest and beginning the RNA isolation procedure, carry out RNase decontamination by wiping down all designated lab benches, dissection tools, and pipettes with an RNase decontamination solution such as RNaseZap® or RNase AWAY®. Prechill cryogenic vials in liquid nitrogen.

3.1 Harvest and Freeze Cartilage from Murine Femoral Heads

1. Obtain wild type mice at 2–3 weeks of age and euthanize by CO_2 inhalation.
2. Dissect hind limbs with surgical scissors and tweezers to expose the femoral head.
3. Peel off cartilage cap from the femoral head using tweezers (*see* **Note 2**) and rinse with sterile PBS to remove blood.
4. Determine the tissue wet weight (for the purpose of this chapter, we pooled cartilage from four mice (i.e., eight cartilage caps) into one prechilled 2 ml cryogenic vial.
5. Snap freeze cartilage in liquid nitrogen for at least 5 min.
6. Proceed directly to tissue pulverization and RNA isolation (Subheading 3.3), or store frozen tissue in −80 °C freezer or in liquid nitrogen.

3.2 Harvest and Freeze Articular Cartilage from Human Osteoarthritic Knee Joint Specimens

On receiving human cartilage specimens from the operating room, it is important that the cartilage is immediately harvested and frozen to maximize RNA yield and integrity. *It is particularly important to avoid storing cartilage tissue in RNA protection reagents such as RNAlater* since this results in dehydration and hardening of the tissue which can subsequently affect RNA yield and integrity [13, 14].

1. Wash cartilage specimens thoroughly with sterile PBS to remove blood.
2. Identify areas of full-thickness articular cartilage to be harvested (*see* **Note 3**).
3. Using a size #10 scalpel blade, cut through the full thickness cartilage and shave tissue off the underlying subchondral bone. If consistency in tissue sampling is required, use a biopsy punch (4–6 mm) to then collect equal-sized full thickness cartilage from the tissue pieces removed from the subchondral bone.
4. Determine the tissue wet weight.

5. Place cartilage into prechilled 2 ml cryogenic vials.
6. Snap freeze cartilage in liquid nitrogen for at least 5 min.
7. Proceed either to tissue pulverization and RNA extraction (Subheading 3.3), or store the tissue in a −80 °C freezer or in liquid nitrogen.

3.3 RNA Purification Using a TRIzol® Reagent and Spin Column Chromatography

1. Place up to 100 mg of snap frozen cartilage tissue in a liquid nitrogen-cooled stainless-steel shaking flask (7 ml) containing a chromium steel grinding ball (*see* **Note 4**).
2. Secure the tissue-containing flask into the holder of the Mikro-dismembrator and activate the machine to shake the flask vigorously for 3 × 30 s (at a maximum shaking frequency of 2000 times/min). A fine tissue powder will be generated.
3. Add 1 ml of chilled TRIzol® which will freeze as it is added (*see* **Note 5**).
4. Once the TRIzol® has thawed, mix thoroughly with a 1 ml pipette to dissolve all powdered tissue in the flask (*see* **Note 6**).
5. Transfer the TRIzol®–tissue mixture to a fresh 1.5 ml microcentrifuge tube and incubate for 5 min at room temperature.
6. Centrifuge at 12,000 × *g* for 10 min at 4 °C to pellet tissue debris.
7. Transfer TRIzol® solution to a fresh 1.5 ml microcentrifuge tube.
8. Add 200 μl of chloroform, vortex briefly (<10 s) and incubate for 10 min at room temperature. Vortex briefly one more time during this 10 min incubation period.
9. Centrifuge at 12,000 × *g* for 10 min at 4 °C.
10. Carefully transfer the upper aqueous phase (~500–550 μl) into a fresh 1.5 ml microcentrifuge tube. Take care not to disrupt the middle interphase layer.
11. With the upper aqueous phase, carry out an additional chloroform extraction by repeating **steps 8–10**.
12. Add 0.6 volumes of isopropanol to the aqueous phase, vortex briefly (<10 s) and incubate for 10 min at room temperature.
13. Centrifuge at 12,000 × *g* for 10 min at 4 °C and discard the supernatant (*see* **Note** 7).
14. Prepare to follow the protocol as described in the Norgen Total RNA Purification Kit manual.
15. Resuspend the centrifuged pellet in 350 μl of lysis buffer RL provided by the Norgen Kit (*see* **Note 8**).
16. Add 200 μl of 100% ethanol to the lysate, briefly vortex to mix and incubate for 10 min at room temperature. This concentration of ethanol is sufficient to isolate small RNAs (<75 nt) as well as mRNAs and long non-coding RNAs.

17. Apply the lysate–ethanol mixture onto a Norgen spin column assembled in a collection tube.
18. Centrifuge at 14,000 × *g* for 1 min at room temperature.
19. Discard the flow-through and reassemble the spin column with the collection tube.
20. Apply 400 μl of Wash Solution A to the column and centrifuge at 14,000 × *g* for 1 min at room temperature.
21. Discard the flow-through and reassemble the spin column in its collection tube.
22. Add 115 μl of DNase I + Enzyme Incubation Buffer to the column and centrifuge at 14,000 × *g* for 1 min at room temperature.
23. Pipette the flow-through in the collection tube back onto the same spin column.
24. Incubate the spin column at 25–30 °C for 15 min.
25. Apply 400 μl of Wash Solution A to the column, centrifuge at 14,000 × *g* for 1 min at room temperature and discard the flow-through.
26. Repeat washing (**step 25**) one more time and discard the flow-through.
27. Centrifuge the spin column (14,000 × *g*) for an additional 2 min at room temperature to dry the resin. Discard the collection tube.
28. Place the column into a fresh 1.5 ml microcentrifuge collection tube.
29. Add 50 μl of Elution Solution A (or RNase/DNase-free water) to the column. Alternatively, add 25–30 μl of elution solution if a more concentrated sample is desired.
30. Centrifuge for 2 min at 200 × *g*.
31. Centrifuge for 1 min at 14,000 × *g*.
32. Analyze the purity and concentration of the RNA sample using a spectrophotometer (NanoDrop™) and determine the RNA integrity/quality using a Bioanalyzer (Agilent).
33. Store the RNA sample at −80 °C until further use in downstream applications.

4 Notes

1. With respect to A_{260}/A_{230} ratios, high A_{230} values in relation to A_{260} is commonly due to contamination from components present in TRIzol® or other lysis buffers. For cartilage tissues, proteoglycan contamination may also affect the A_{230} values.

We cannot rule out that traces of proteoglycan remain within the samples with our preferred method, but we suspect that the majority have been removed and that the variations in A_{260}/A_{230} values we see between the murine and human samples tested here is due to some carryover of TRIzol® or Norgen Kit lysis buffer components. Of interest is a statement made in a Qiagen Newsletter from 2010 that concentrations of guanidine thiocyanate of up to 100 mM in an RNA sample do not compromise the reliability of downstream applications. https://www.qiagen.com/us/resources/faq?id=c59936fb-4f1e-4191-9c16-ff083cb24574&lang=en.

2. To assist in harvesting murine femoral head cartilage, utilizing a dissecting microscope or a magnifying glass lamp is advised to aid in identifying the white cartilage "cap." Once identified, hold the mouse hind limb with your gloved hand or tweezers and then carefully peel off the cartilage cap with fine tweezers. Use of a magnification device will certainly be critical if the goal is to isolate mature murine articular cartilage from knee or hip joints. Such murine tissue is extremely thin which renders the dissection procedure quite challenging. Also be aware that pooling mature murine articular cartilage tissue from a number of limbs will be necessary to obtain sufficient amounts of RNA for downstream analysis.
3. It is important to thoroughly inspect OA cartilage specimens to determine suitable areas that should be removed for sampling. Areas of white-looking, full-thickness articular cartilage are quite obvious and are suitable for collection. Avoid areas of partial thickness cartilage that clearly look degraded and will most likely correspond to areas of acellular tissue or tissue containing dead cells. Also avoid areas where the tissue has a yellow discoloration as well as regions of "neo" cartilage that are usually visible around the periphery of the joint.
4. For this procedure, 100 mg of tissue is the maximum amount recommended. Attempting to extract RNA from more tissue will likely reduce RNA yield. If purifying RNA from more than 100 mg of tissue, then either scale up the volume of TRIzol® accordingly, or divide the tissue into equal portions of 100 mg or less and combine resulting aliquots of RNA [14].
5. Perform all steps involving TRIzol® in the fume hood.
6. An optional step to ensure powdered tissue is fully dissolved in TRIzol® is to pass this solution 10–20 times through an 18G–22G syringe needle. If the solution foams, wait for the foam to settle before proceeding to the next step.
7. At this stage, a white RNA pellet is often visible by eye. However, even if no pellet is detectable, it is advised to continue with the procedure as it is likely that RNA is present, albeit at low levels.

8. Even though cell lysis has already been achieved earlier in the procedure with addition of TRIzol®, we add Norgen Kit's lysis buffer at this stage to ensure buffer compatibility with the spin column resin.

Acknowledgments

This work was supported by NIH grant R01 AR064191 (to A.M). We thank staff members of the Genome Technology and Access Center at Washington University for assistance with Bioanalyzer analysis. We also thank Dr. Ryan Nunley M.D. in the Department of Orthopaedic Surgery at Washington University for providing us with human osteoarthritic articular cartilage specimens.

References

1. Kuettner KE, Aydelotte MB, Thonar EJ (1991) Articular cartilage matrix and structure: a minireview. J Rheumatol Suppl 27:46–48
2. Hardingham T, Bayliss M (1990) Proteoglycans of articular cartilage: changes in aging and in joint disease. Semin Arthritis Rheum 20:12–33
3. Kerin A, Patwari P, Kuettner K, Cole A, Grodzinsky A (2002) Molecular basis of osteoarthritis: biomechanical aspects. Cell Mol Life Sci 59:27–35
4. Yang M, Zhang L, Gibson GJ (2015) Chondrocyte miRNAs 221 and 483-5p respond to loss of matrix interaction by modulating proliferation and matrix synthesis. Connect Tissue Res 56:236–243
5. Hayman DM, Blumberg TJ, Scott CC, Athanasiou KA (2006) The effects of isolation on chondrocyte gene expression. Tissue Eng 12:2573–2581
6. Adams ME, Huang DQ, Yao LY, Sandell LJ (1992) Extraction and isolation of mRNA from adult articular cartilage. Anal Biochem 202:89–95
7. Smale G, Sasse J (1992) RNA isolation from cartilage using density gradient centrifugation in cesium trifluoroacetate: an RNA preparation technique effective in the presence of high proteoglycan content. Anal Biochem 203:352–356
8. Mallein-Gerin F, Gouttenoire J (2004) RNA extraction from cartilage. Methods Mol Med 100:101–104
9. Ghawana S, Paul A, Kumar H, Kumar A, Singh H, Bhardwaj PK, Rani A, Singh RS, Raizada J, Singh K, Kumar S (2011) An RNA isolation system for plant tissues rich in secondary metabolites. BMC Res Notes 4:85
10. McKenna LA, Gehrsitz A, Soder S, Eger W, Kirchner T, Aigner T (2000) Effective isolation of high-quality total RNA from human adult articular cartilage. Anal Biochem 286:80–85
11. Gehrsitz A, McKenna LA, Soder S, Kirchner T, Aigner T (2001) Isolation of RNA from small human articular cartilage specimens allows quantification of mRNA expression levels in local articular cartilage defects. J Orthop Res 19:478–481
12. Miyaki S, Nakasa T, Otsuki S, Grogan SP, Higashiyama R, Inoue A, Kato Y, Sato T, Lotz MK, Asahara H (2009) MicroRNA-140 is expressed in differentiated human articular chondrocytes and modulates interleukin-1 responses. Arthritis Rheum 60:2723–2730
13. Ruettger A, Neumann S, Wiederanders B, Huber R (2010) Comparison of different methods for preparation and characterization of total RNA from cartilage samples to uncover osteoarthritis in vivo. BMC Res Notes 3:7
14. Le Bleu HK, Kamal FA, Kelly M, Ketz JP, Zuscik MJ, Elbarbary RA (2017) Extraction of high-quality RNA from human articular cartilage. Anal Biochem 518:134–138

Chapter 10

Small Nucleolar RNA Expression Profiling in Cartilage

Mandy J. Peffers, Andy Cremers, and Tim J. M. Welting

Abstract

Osteoarthritis (OA) presents as a change in the articular chondrocyte phenotype. The origin of the phenotype change is poorly understood. Small nucleolar RNAs (snoRNAs) direct chemical modification of other RNA substrates and are involved in endoribonucleolytic pre-rRNA processing. They have thereby a role by fine-tuning spliceosome and ribosome function and can thus accommodate changing requirements for protein synthesis in OA. Here we describe both targeted and global methods for snoRNA isolation and quantification from whole cartilage.

Key words snoRNA, Cartilage, Osteoarthritis, snoRNA-Seq qRT-PCR, Primer design

1 Introduction

Cartilage is a part of the skeletal system that, in articular joints, acts like as a shock absorber at the contact points of opposing long bones. Articular cartilage consists of a single cell type, the chondrocyte, which is embedded in an abundant extracellular matrix (ECM) consisting mainly of collagen type II and the proteoglycan aggrecan. Osteoarthritis (OA) is a whole-joint disease that includes gradual degradation of articular cartilage. There are multifactorial risk factors that trigger OA is multifactorial, including aging [1].

Many independent studies have demonstrated that small nucleolar RNAs (snoRNAs), in addition to other noncoding RNAs, serve as short regulatory RNAs. SnoRNA levels in human tissues, cells, and blood present as promising targets for diagnostics and treatments of human pathologies (reviewed [2]). SnoRNAs direct chemical modification of other RNA substrates and are mainly involved in endoribonucleolytic pre-rRNA processing. The post-transcriptional 2'*O*-ribose methylation and pseudouridylation carried out by Box C/D and Box H/ACA snoRNAs (respectively) fine-tune spliceosome and ribosome function, accommodating changing requirements in protein synthesis during health and disease. The control of snoRNA levels may be pivotal in regulating the

Tariq M. Haqqi and Véronique Lefebvre (eds.), *Chondrocytes: Methods and Protocols*, Methods in Molecular Biology, vol. 2245, https://doi.org/10.1007/978-1-0716-1119-7_10,

transcriptional and translational capacity of chondrocytes. To ensure continuous extracellular matrix (ECM) turnover, it is essential for a chondrocyte to control the number and quality of its ribosomes. Pathways controlling ribosome activity have previously been described in the regulation of chondrocyte homeostasis [3, 4]. We have previously identified age- and disease-related changes in snoRNA expression in cartilage, chondrocytes, tendon, mesenchymal stem cells, serum and synovial fluid [5–12].

The purpose of this chapter is to describe and discuss the measurement of snoRNAs in cartilage tissue. We have used a combination of nonbiased small RNA-seq (which we term snoRNA-seq due to the implementation of a 5′ decapping prior to library preparation) and targeted qRT-PCR. As we believe best practice is to assess these molecules in whole cartilage, we have focused on three main areas: selection of cartilage samples, techniques for the production of poly(A) cDNA without the use of bespoke kits, and design of snoRNA primers to enable the researcher to design their own primers.

Thus we describe in the following sections a workflow used for detecting snoRNAs from cartilage either using a targeted approach of qRT-PCR or using an unbiased approach of snoRNA-Seq. While our cartilage RNA extraction method may be standard in most cartilage laboratories, our method used to produce poly (A) cDNA is adapted from the original paper describing detection and quantification of small RNAs by the Yan group [4]. Furthermore, while others have undertaken snoRNA-Seq on a variety of musculoskeletal samples [13–15] we have always added an additional step to our standard snoRNA-seq workflow prior to library preparation in order to enable unbiased sequencing by removal of the 5′ caps found on some snoRNAs [5, 6, 16].

To summarize, determination of snoRNAs in cartilage can yield important information. Resulting data may lead to a broader understanding of the relationships between cartilage ageing, osteoarthritis, and the chondrocyte phenotype.

2 Materials

All reagents are prepared in RNase-free DNase-free MilliQ water, unless stated otherwise. All steps are performed at room temperature unless stated otherwise.

2.1 Collection of Cartilage Biopsies for RNA Extraction and Histological Grading (See Note 1)

1. Capped jars with 0.9% NaCl, autoclaved.
2. Instrument set: 1× knife handle with size 20 scalpel blade, 1× surgical tweezers, 1× surgical tweezers 1/2 teeth, autoclaved or sterilized in alternative manner.
3. 15-cm petri dish.

4. 0.9% NaCl solution, autoclaved.
5. Disposable 8-mm biopsy punch.
6. RNase-free sterile 2-ml Eppendorf tubes.
7. Liquid nitrogen or RNAlater™.
8. 3.7% formaldehyde.
9. 0.5 M ethylenediaminetetraacetic acid (EDTA) pH 8.0.
10. 70% ethanol, molecular biology grade.
11. Histology embedding cassettes.

2.2 Micro-dismembrator Tissue Homogenization for RNA Extraction (See Notes 2 and 3)

1. Micro-dismembrator (B. Braun) including metal capsule, metal ball, metal lid, and rubber ring.
2. Clean jars for decontamination.
3. Decontamination solution (0.1 M NaOH, 10% sodium dodecyl sulphate (SDS)).
4. DEPC water (Sigma-Aldrich).
5. 70% ethanol, molecular biology grade.
6. Liquid nitrogen.
7. RNase-free sterile 2-ml Eppendorf tubes.
8. Microbalance.
9. RNase ZAP (Sigma-Aldrich).
10. Large tweezers.
11. Surgical artery forceps with scissor handle.
12. mirVana™ Lysis Binding Buffer (Applied Biosystems).

2.3 RNA Isolation from Cartilage and Subsequent Production of Poly(a) RNA

1. mirVana Total RNA Isolation Kit (Invitrogen Life Technologies).
2. 100% Ethanol, molecular biology grade.
3. Acid Phenol Chloroform (Invitrogen Life Technologies).
4. RNase-free sterile 0.2-ml thin wall PCR tubes.
5. RNasin Ribonuclease Inhibitor (Promega).
6. Poly(A) Polymerase Reaction Buffer 10x concentrated (New England Biolabs).
7. *E. coli* Poly(A) Polymerase 5.0 U/μl (New England Biolabs).
8. Adenosine triphosphate (ATP) 10 mM (New England Biolabs).

2.4 Poly(A) RNA Isolation

1. MirVana™ Total RNA Isolation Kit (Invitrogen Life Technologies).
2. 100% ethanol, molecular biology grade.
3. Acid-phenol–chloroform (Invitrogen Life Technologies).

2.5 Poly(A) cDNA Synthesis

1. RNase-free sterile 0.2 ml thin wall PCR tubes.
2. RTQ primer, CGA-ATT-CTA-GAG-CTC-GAG-GCA-GGC-GAC-ATG-GCT-GGC-TAG-TTA-AGC-TTG-GTA-CCG-AGC-TCGGAT-CCA-CTA-GTC-CTT-TTT-TTT-TTT-TTT-TTT-TTT-TTT-TTV-N
3. RNase-free DNase-free water.
4. M-MLV reverse transcriptase, 200 U/μl (Promega).
5. RNasin Ribonuclease. Inhibitor (Promega)
6. dNTP set, deoxynucleotide triphosphates dATP, dCTP, dGTP, and dTTP, 100 mM (Tebu-Bio).
7. M-MLV Reverse Transcriptase Buffer, 5× concentrated (Promega).
8. RNase H, Ribonuclease H recombinant (Bioke).
9. RNase H Reaction Buffer, 10× concentrated (Bioke).

2.6 SnoRNA Primer Design

1. Website OligoCalc; Basic; http://biotools.nubic.northwestern.edu/OligoCalc.html.
2. *See* Table 1.

2.7 qRT-PCR

1. MESA GREEN qPCR MasterMix Plus for SYBR Green.
2. SnoRNA-specific forward primer.
3. Melt temperature-matched universal reverse primer (REV-UNI-XX).
4. RNase-free DNase-free water.

2.8 CAP-CLIP Pyrophosphatase Pretreatment Prior to snoRNA-seq Library Preparation

1. Cap-Clip™ Acid Pyrophosphatase (Cambio or similar) kit containing.
2. Cap-Clip™ Acid Pyrophosphatase, 5 U/μl.
3. 10× Cap-Clip™ Acid Pyrophosphatase Reaction Buffer.
4. Ampure RNA XP beads.
5. RNase-free water.
6. Molecular biology grade ethanol.
7. Molecular biology grade isopropanol.
8. Magnet.

3 Methods

3.1 Collection of Cartilage Explants for RNA Extraction and Histological Grading

1. Articular cartilage surgical remnants are collected freshly during total knee arthroplasty surgery. These remnants are part of the waste material from the surgery and their use for research should be approved by the appropriate local medical ethics body. Collect cartilage parts at the operation room in the sterile

Table 1
(A) RTQ primer with poly T part, (B) Universal reverse primer, and (C) Reference genes (ncRNA-specific forward primers)

(A)				
Primer	**Information**	**Sequence (5′–3′)**	**Length (nt)**	
RTQ primer	Poly(T) primer for reverse transcription	CGA-ATT-CTA-GAG-CTC-GAG-GCA-GGC-GAC-ATG-GCT-GGC-TAG-TTA-AGC-TTG-GTA-CCG-AGC-TCG-GAT-CCA-CTA-GTC-CTT-TTT-TTT-TTT-TTT-TTT-TTT-TTT-TTV-N	100	
(B)				
Primer	*Information*	*Sequence (5′–3′)*	*Length (nt)*	*Tm OligoCalc basic*
REV-UNI-50	Reverse primer	TAG-TTA-AGC-TTG-GTA-CCG-AG	20	49.7 °C
REV-UNI-55	Reverse primer	AAT-TCT-AGA-GCT-CGA-GGC-AGG	21	54.3 °C
REV-UNI-60	Reverse primer	CGA-ATT-CTA-GAG-CTC-GAG-GCA-GG	23	58.8 °C
REV-UNI-65	Reverse primer	CGA-ATT-CTA-GAG-CTC-GAG-GCA-GGC-GAC	27	64.3 °C
REV-UNI-70	Reverse primer	CTA-GAG-CTC-GAG-GCA-GGC-GAC-ATG-GCT-GGC	30	69.8 °C
(C)				
Primer	*With REV-UNI primer*	*Sequence (5′–3′)*	*Bases*	*Tm OligoCalc basic*
Hs 5S rRNA	REV-UNI-60	CCT-GGG-AAT-ACC-GGG-TGC-TGT-AG	23	60.6 °C
Hs U2 snRNA	REV-UNI-60	TGG-TAT-TGC-AGT-ACC-TCC-AGG-AAC-G	25	59.3 °C

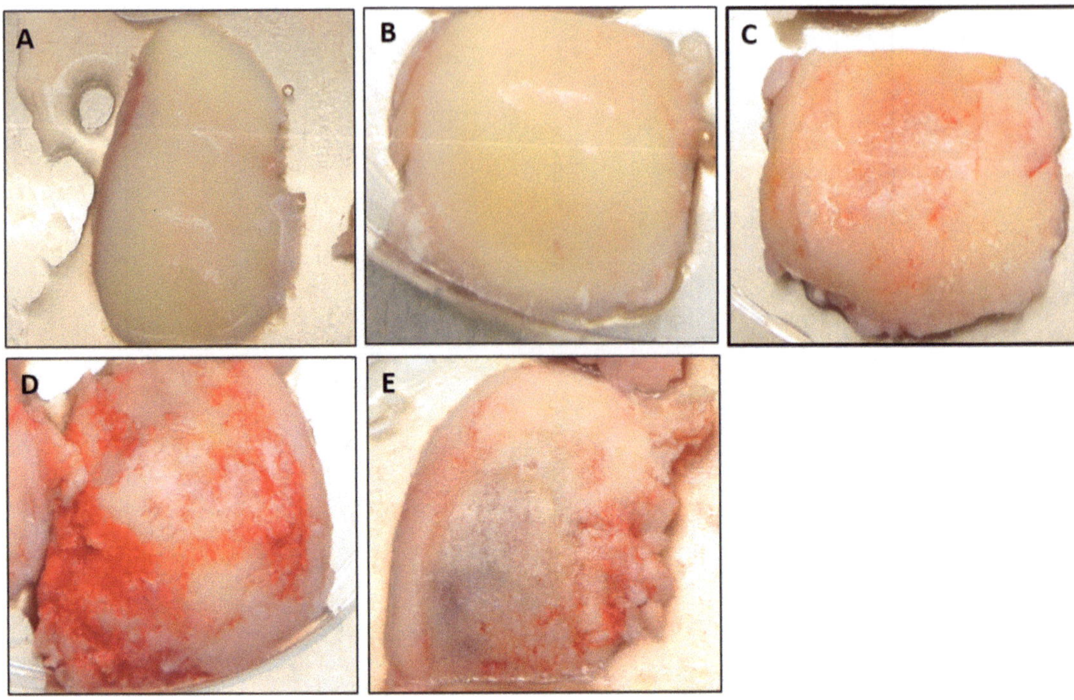

Fig. 1 Photographs of human knee cartilage removed following total knee arthroplasty. Images show examples of the grades as derived from the Outerbridge Classification System [17]. (**a**) Protected; this area of cartilage is normal or near normal. (**b–d**) Unprotected; cartilage visibly affected by osteoarthritis. The level of severity is graded grossly by eye. If a lesion is present, biopsies are taken from the side immediately adjacent. (**b**). Grade 1; no lesions visible but cartilage is roughened. (**c**) Grade 2; clear lesions visible but some full-thickness cartilage remains. (**d**) Grade 3; clearly visible lesion exposing the subchondral bone. (**e**) Grade 4; large areas of exposed subchondral bone

capped jars with 0.9% NaCl and transport with lid closed to the laboratory for further processing.

2. Using the instrument set, take out cartilage parts from the jar and anatomically order the cartilage parts in a 15-cm petri dish in a laminar flow cabinet. If necessary, wash blood parts away with 0.9% NaCl. Do not allow tissue to dry out.

3. Depending on the research question, one can take samples from protected (not affected by OA) and/or contralateral unprotected (OA affected) cartilage. Define protected and unprotected regions (Fig. 1). For objective reviewing afterward, take a photograph and visually grade the osteoarthritic stage (Fig. 1).

4. Remove all unused cartilage parts from the petri dish and continue with the selected part(s). Take the 8-mm dermal biopsy punch and push punches in the selected unprotected and protected regions (Fig. 2) until subchondral bone has been reached. Take at least two biopsies for RNA isolation. It is a

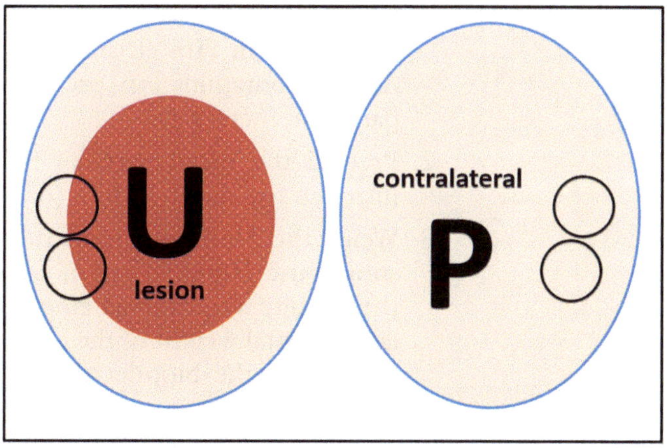

Fig. 2 Diagram of part of the patient detail form to mark the locations of cartilage sampling following total knee arthroplasty. U; unprotected (usually the medial condyle), P; protected (usually the lateral condyle). Black circles mark where the biopsies are taken using 8-mm punch biopsies

consideration to also take biopsies of each sampled region for histological OA grading. Use the scalpel knife to cut all the biopsies from the subchondral bone. Allow the scalpel knife to cut in a horizontal way, following the border between the tough subchondral bone and the softer cartilage layer.

5. For RNA isolation, collect the cut biopsies in a 2-ml Eppendorf tube and snap freeze in liquid nitrogen for further processing or storage at −80 °C. Alternatively, for RNA isolation, biopsies may be collected into RNAlater™ according to manufacturer's instructions. Continue to Subheading 3.2.
6. For histology, collect the biopsies in a 2-ml Eppendorf tube with 1500 μl 3.7% formaldehyde. Incubate overnight at 4 °C while rotating. After 24 h, replace by 0.5 M EDTA pH 8.0 for decalcification, and refresh after 7 days. After 14 days, wash in 70% ethanol and embed the biopsies in paraffin using embedding cassettes. The recommended device for paraffin embedding is the Sakura Tissue-Tek VIP 2000. Follow standard histology protocols for further sectioning, staining, and OA grading.
7. Continue with Subheading 3.2; tissue homogenization for RNA extraction.

3.2 Micro-dismembrator Tissue Homogenization for RNA Extraction [18]

1. For RNA isolation, the collected cartilage biopsies must be pulverized by disintegrating the biopsies. All capsule parts of the micro-dismembrator (metal capsule, metal ball, metal lid, and rubber ring) undergo a short procedure to reduce risk for RNase contamination. This same procedure should be repeated for every second time a sample is processed. Procedure: 10 min

in decontamination solution, 3 × 1 min in DEPC water, and 3 × 1 min in 70% ethanol. Allow to air-dry in a fume hood. After decontamination, prevent any contact with inner capsule parts and metal ball.

2. Prechill the metal parts of the assembled capsule in liquid nitrogen for about 10 min (or until bubbling stops).
3. Weigh the 2-ml Eppendorf tube with biopsies. Generally, one should use 100–250 mg of cartilage biopsy tissue. Determine the average weight of an empty 2-ml Eppendorf tube to calculate the total weight of the cartilage biopsies for disintegration. Make sure the biopsies do not thaw during this procedure; work fast.
4. Transfer the prechilled capsule to an RNase ZAP™-cleaned surface in the fume hood and open the lid. Transfer the frozen biopsy into the capsule by decanting, and close the lid including metal ball and rubber ring. If the biopsy does not drop out of the tube, use an RNaseZAP™-cleaned tweezers for transferring the biopsy (*see* **Note 4**).
5. Chill the capsule-biopsy combination again in liquid nitrogen until bubbling stops. Use surgical artery forceps with scissor handle to hold the capsule and avoid the combination falling apart.
6. Clamp the closed capsule into the micro-dismembrator (work fast to avoid warming-up again) and disintegrate the sample for 1 min at 2600 rpm on a stable surface.
7. Weigh the now empty Eppendorf tube to determine the exact weight of the corresponding biopsy.
8. After disintegration, carefully transfer the capsule to the RNase ZAP™ cleaned surface in the fume hood. Open the capsule with care. Be aware that the pulverized cartilage will also be inside the lid, attached to the metal ball and rubber ring.
9. Add 500–2500 μl MirVana Lysis Binding Buffer™ (10× volume Lysis Binding Buffer per tissue mass is required). The cartilage powder will still be frozen. Since the buffer will freeze, divide the buffer volume between capsule, metal ball, rubber ring and the lid, making sure to reach all the powder. Allow to thaw at room temperature and pipette the buffer up-and-down to collect any remaining powder and to aid sample lysis (*see* **Note 5**).
10. Collect the complete buffer/sample volume in a new RNase-free 2-ml (or 5-ml) Eppendorf tube.
11. Continue with Subheading 3.3, RNA isolation from cartilage.

3.3 RNA Isolation from Cartilage for Subsequent Poly-A RNA [19]

1. After dissolving the pulverized cartilage in 500–2500 μl Lysis Binding Buffer, incubate for 10 min on ice before proceeding.
2. Follow the manufacturer's protocol of the MirVana miRNA Isolation Kit, selecting the correct instruction for RNA extraction from tissue. **Steps 3–9** are identical to this manufacturer's protocol. First, prepare the kit solutions before starting the isolation.
3. Add 1/10 volume of miRNA Homogenate Additive and mix well by vortexing. Incubate for 10 min on ice. Use the fume hood to add 1 volume of acid-phenol–chloroform (1:1 to Lysis Binding Buffer volume) and mix well by vortexing.
4. Centrifuge for 30 min at 15,000 rpm (room temperature). Carefully (without disturbing the phase separation) transfer the upper aqueous phase into a new RNase-free 2-ml (or 5-ml) Eppendorf tube and take note of the transferred volume. Meanwhile, preheat the Elution Solution at 95 °C for at least 5 min (*see* **Note 6**).
5. Add 5/4 volume of 100% ethanol to the transferred aqueous phase and mix well by vortexing. Place a filter cartridge in a new MirVana collection tube.
6. Transfer a maximum of 700 μl of the sample–ethanol mix to the filter cartridge. Centrifuge for 30 s at 10,000 rpm (room temperature) and discard the flow through. If the total volume exceeds 700 μl, repeat this step until the complete volume has passed the filter.
7. Wash the filter by adding 700 μl of miRNA Wash Solution 1 and centrifuge for 30 s at 10,000 rpm (room temperature). Discard the flow-through. Repeat the same procedure twice by using 700 μl miRNA Wash Solution 2/3.
8. To remove any remaining liquid before eluting, centrifuge one more time for 30 s at 10,000 rpm without any solution on top of the filter. Replace the collection tube.
9. Quickly add 100 μl of the 95 °C-heated Elution Solution to the filter. Lower the volume (but $\geq$60 μl) if a low yield is expected. Be aware that the Eppendorf tube with the heated Elution Solution may be pressurized. Centrifuge for 30 s at 10,000 rpm to elute the RNA from the filter. Measure the RNA concentration and quality using spectroscopy. A recommended device is the Nanodrop One.
10. If the RNA concentration permits it, use up to 1000 ng total RNA for the poly(A) RNA synthesis. The maximum volume is 36 μl, so the RNA concentration is $\geq$28 ng/μl.
11. Add together in an RNase-free 0.2-ml thin wall PCR tube:
    36–*X* μl RNase free DNase-free H_2O

11 μl mix of

1 μl RNasin Ribonuclease Inhibitor, 20 U/μl

5 μl Poly(A) Polymerase Reaction Buffer, 10× concentrated

5 μl ATP, 10 mM

X μl RNA, up to 1000 ng

3 μl *E. coli* poly(A) polymerase, 0.5 U/μl (prepare a working stock to reach this concentration)

12. Polyadenylate the RNA sample in a thermocycler at 37 °C for 60 min. The total sample volume at this stage is 50 μl poly (A) RNA. A recommended thermocycler is the Biometra TRIO thermocycler.
13. Continue with Subheading 3.4, poly(A) RNA isolation.

3.4 Poly(A) RNA Isolation

1. Use the complete volume of produced poly(A) RNA (50 μl), and add 500 μl of MirVana Lysis Binding Buffer. Mix well by vortexing.
2. Follow the MirVana miRNA Isolation Kit, by starting with the addition of 1/10 volume miRNA Homogenate Additive (55 μl) and mix well by vortexing. The remainder of the procedure for Subheading 3.4 is identical to Subheading 3.3, **steps 3–9**.
3. Continue with Subheading 3.5, poly(A) cDNA synthesis.

3.5 Poly(A) cDNA Synthesis

1. If the yield permits it, use 400 ng poly(A) RNA for the poly (A) cDNA synthesis. The maximum volume is 25 μl, so the [poly(A) RNA] is ≥16 ng/μl.
2. Prepare a stock of the RTQ primer to use for the poly (A) cDNA synthesis. The concentration of this stock depends on the quantity of input poly(A) RNA. The poly(A) RNA–RTQ primer ratio must be 2:1 (w/w). In addition, 1 μl of RTQ primer volume must be used per sample. Therefore, the maximum concentration of the RTQ primer is 200 ng/μl.
3. Add together in an RNase-free 0.2-ml thin-wall PCR Eppendorf tube:

   25–*X* μl RNase-free DNase-free H_2O

   X μl poly(A) RNA, 400 ng

   1 μl RTQ primer (prepare a stock solution based on amount of input RNA in RNase-free DNase-free water). Anneal the RTQ primer to the poly(A) RNA in a thermocycler using a two-step reaction. 10 min at 65 °C and 20 min at 4 °C. The total sample volume after this step is 26 μl. A recommended device is the Biometra TRIO thermocycler.
4. Add the following mix to the 26 μl from stage 3:

14 μl mix of:

1 μl M-MLV, 200 U/μl

1 μl RNasin, 20 U/μl

2 μl dNTP's, 10 mM of each dNTP (40 mM combined dNTPs)

8 μl RT buffer, 5× concentrated

2 μl RNase-free DNase-free water

Reverse transcribe the poly(A) cDNA in a thermocycler in a new two-step reaction. First, heat for 60 min at 50 °C. This step is at a high temperature to have a more efficient unfolding and melting of the structured RNAs. Second, incubate for 15 min at 70 °C. The total sample volume after this step is 40 μl.

5. Add to the 40 μl 3 μl RNase H, 0.5 U/μl (prepare a stock in RNase H Reaction Buffer, 1×). Finalize the poly(A) cDNA synthesis in a thermocycler for 60 min at 37 °C.
6. Continue with Subheading 3.6, snoRNA primer design.

3.6 SnoRNA Primer Design

1. Acquire the sequence of the mature snoRNA of interest using the public repositories.
2. Scan the sequence manually for a potential primer sequence of which the melting temperature (calculated using OligoCalc) is between 50 °C and 65 °C (preferably around 60 °C) and with a length of 20–30 nucleotides. If possible, choose sequence regions with high sequence variety to circumvent cross-reactivity with other RNA species as much as possible. This may be more challenging for box H/ACA snoRNAs than for box C/D snoRNAs, since box H/ACA snoRNAs have higher A/U content than box C/D snoRNAs. BLAST the selected primer sequence against the organism's (in this case *Homo sapiens*) transcriptome and make sure that the primer is not complementary to other RNA species. As a rule of thumb, sequence complementarity between the primer and other transcripts should be less than 80%. In addition, complementarity at the 3′ end of the primer with a potential other transcript should be avoided.
3. When the primer sequence meets the above criteria, it can be ordered in sense orientation and used in Subheading 3.7, qRT-PCR.

3.7 qRT-PCR

1. For use in qPCR reaction, the poly(A) cDNA product must be diluted to 400 ng/43 μl. The optimal and final concentration for RT-qPCR is 2 ng/μl. Then, mix a selection of representative samples to produce a standard series comprising 5×, 16×, 80×, 400×, and 2000× dilutions.

2. To determine the optimal annealing temperature of each unique forward primer (snoRNA of interest) with a reverse primer, a qPCR gradient needs to be run. For each unique forward primer, select the reverse primer (from Table 1 provided in Subheading 2.6) whose melting temperature most closely approaches the melting temperature of the forward primer. This is calculated using OligoCalc: http://biotools.nubic.northwestern.edu/OligoCalc.html.

   There are five different REV-UNI reverse primers, starting at a melting temperature of 50 °C, and each time increasing every 5 °C (50, 55, 60, 65, and 70 °C).

   For human samples we suggest the use of the following noncoding RNAs as a reference for qPCR normalization:

   (a) 5S rRNA, with REV-UNI-60 at an annealing temperature of 67.0 °C.

   (b) U2 snRNA, with REV-UNI-60 at an annealing temperature of 63.3 °C.

      Apart from snoRNAs, other noncoding RNAs will be reverse-transcribed using this method and can thus be used for study or as a reference for qPCR normalization. The primers for human 5S rRNA and U2 snRNA have been validated in our lab and PCR amplicons have been sequence-verified.

3. Real-time quantitative PCR (RT-qPCR) can be performed using a laboratory's preferred qPCR master mix, protocol, and qPCR machine. A protocol working in the author's labs will be explained below. RT-qPCR is performed using MESA GREEN qPCR MasterMix Plus for SYBR Green in 96-well optical plates. Forward and reverse primers should be used at 300 nM each. Basic temperature protocol: initial denaturation at 95 °C for 10 min, followed by 50 cycles of DNA amplification (denaturation for 15 s at 95 °C and annealing for 1 min at an empirically selected annealing temperature) followed by a dissociation curve. For initial use and optimisation of a novel snoRNA-specific forward primer, an annealing temperature gradient should be run to find the optimal annealing temperature of the forward–reverse primer combination. Make a qPCR plate setup with the same dilution set on each row. Include a nontemplate control. To run an annealing temperature gradient, the recommended instrument is a Bio-Rad CFX96 Optical Reaction Module with C1000 Real-Time PCR. Start with a wide-range annealing temperature gradient, such as 44–66 °C. The software automatically divides intermediate temperatures per row. Run the RT-qPCR.

4. Analyze the run report and rate each row for the following parameters:

(a) Identical melting peak for all standard series dilutions.

(b) Quality of the standard curve (the optimal slope is −3.3).

(c) Width of the standard curve range (a broader range provides greater detectability).

(d) C_t value of the nontemplate control relative to the standard curve (distance should be far enough to exclude aspecificity of the PCR reaction).

Based on these criteria, select the best annealing temperature to continue. Optional: (1) Run qPCR samples on an appropriate agarose gel to confirm the anticipated PCR amplicon size; (2) sequence the amplicon for final confirmation of the qPCR specificity for the respective snoRNA.

5. If further optimization is required, repeat **step 3** with a narrower annealing temperature gradient, and with the optimal annealing temperature from **step 4** as a starting point. For example, if 57.8 °C was selected to be optimal in **step 4**, choose 53–61 °C as a new range. Temperature jumps between the rows will become much smaller now. Repeat **step 4** and select the final optimal annealing temperature based on this result.
6. Run the RT-qPCR with optimized annealing temperature and take along a reference noncoding RNA for qPCR normalization (we suggest 5S rRNA and U2 snRNA). RT-qPCR can be run using either the standard curve method or the delta–delta C_t method.

3.8 CAP-CLIP Pyrophosphatase Pretreatment Prior to snoRNA-Seq Library Preparation for Hydrolysis of the Pyrophosphate Bonds of the 5′ Cap of snoRNAs (See Note Note 7)

1. This step is undertaken prior to library preparation.
2. Up to 1 μg total RNA from the above isolation in Subheading 3.2 is incubated for 1 h at 37 °C with 5 U CAP-CLIP in 1× reaction buffer.
3. Clean up with Ampure RNA XP beads. Bring the sample volume up to 50 μl with RNase-free water.
4. Add 90 μl Ampure RNA XP beads and 270 μl isopropanol. Leave this mixture for 5 min at room temperature.
5. Pull back the beads with a magnet and remove the supernatant. Wash the beads with 300 μl 85% ethanol twice and remove all traces of ethanol.
6. Resuspend the beads in 6 μl RNase-free water and leave for 2 min at room temperature. Pull back the beads and recover the liquid into a clean tube for the 3′ ligation step.
7. Proceed with material into library preparation using manufacturer's recommendation (*see* **Note 8**).

4 Notes

1. This protocol describes the procedure for taking biopsies of human knee cartilage. To investigate the quality (we also score by the Kellgren & Lawrence score) of the cartilage, biopsies needs to be taken in the PROTECTED (P) and UNPROTECTED (U) area. These regions must be contralateral to each other.
2. Several precautions should be taken prior to sample extraction. Bench surfaces should be cleaned. Gloves should be worn at all times when taking Eppendorf tubes from plastic bags and handling pipette tips, and to avoid contaminating samples with skin nucleases, proteases, and oils. Lab equipment should be cleaned in soapy water before use. The dismembrator should be cleaned and dried before use to prevent sample contamination. Safety glasses should be worn while handling liquid nitrogen. Ear protection should be worn while disintegration.
3. Micro-disintegration for tissue homogenization of whole cartilage to isolate RNA. This technique is based on the enlargement of the surface area by powdering the cartilage, which results in faster RNA isolation. With different shaking capsules and metal grinding balls available, the micro-dismembrator can be adapted to each individual application.
4. Tape in your thumb and forefinger and wear double purple gloves to prevent freezing of your hands. Work fast throughout the procedure as homogenization requires a deep-frozen sample and capsule. Once thawing starts, the sample will not powder properly.
5. For other RNA isolation methods add 750 μl TRIzol.
6. We use water as because chemicals in the kit elution buffer may interfere with the preparation of libraries for snoRNA-Seq.
7. Cap-Clip™ Acid Pyrophosphatase is a replacement for the Tobacco Acid Pyrophosphatase (TAP), which is no longer available. This plant-derived enzyme hydrolyzes numerous pyrophosphate bonds, including those necessary to convert the cap structures (decapping) of snoRNAs into 5′-monophosphate RNAs. We incorporate this step into our standard workflow for snoRNA-Seq to enable unbiased sequencing. This 5′-cap-removal step is applied before library preparation.
8. For snoRNA-Seq we use a recommended sequencing facility. As standard we submit total RNA to the facility which they assess RNA for quality using RNA integrity measurements. While it is possible to prepare libraries yourself, we feel that using the services of a facility that regularly carries out these methodologies is better in order to reduce operator bias.

References

1. Li Y et al (2013) The age-related changes in cartilage and osteoarthritis. Biomed Res Int 2013:916530
2. Stepanov GA et al (2015) Regulatory role of small nucleolar RNAs in human diseases. Biomed Res Int 2015:206849
3. Pirog KA et al (2014) Abnormal chondrocyte apoptosis in the cartilage growth plate is influenced by genetic background and deletion of CHOP in a targeted mouse model of pseudoachondroplasia. PLoS One 9(2):e85145
4. Yamabe S et al (2013) Intracellular accumulation of advanced glycation end products induces apoptosis via endoplasmic reticulum stress in chondrocytes. FEBS J 280 (7):1617–1629
5. Steinbusch MM et al (2017) Serum snoRNAs as biomarkers for joint ageing and post traumatic osteoarthritis. Sci Rep 7:43558
6. Peffers MJ et al (2016) Age-related changes in mesenchymal stem cells identified using a multi-omics approach. Eur Cell Mater 31:136–159
7. Pease LI et al (2017) Cross platform analysis of transcriptomic data identifies ageing has distinct and opposite effects on tendon in males and females. Sci Rep 7(1):14443
8. Peffers MJ et al (2015) Transcriptome analysis of ageing in uninjured human Achilles tendon. Arthritis Res Ther 17:33
9. Peffers MJ, Liu X, Clegg PD (2014) Transcriptomic profiling of cartilage ageing. Genom Data 2:27–28
10. Peffers MJ et al (2018) A role for the snoRNA U3 in the altered translational capacity of ageing and osteoarthritic chondrocytes. Osteoarthritis Cartilage 26(Supplement 1):S45–S46
11. Peffers MJ et al (2018) snoRNA signatures in cartilage ageing and osteoarthritis. Osteoarthritis Cartilage 26(Supplement 1):S164
12. Peffers MJ et al (2018) Small non coding RNAs as biomarkers of early osteoarthritis. Equine Vet J 50(552):S45
13. Coutinho de Almeida R et al (2019) RNA sequencing data integration reveals an miRNA interactome of osteoarthritis cartilage. Ann Rheum Dis 78(2):270–277
14. Chen YJ et al (2018) Systematic analysis of transcriptomic profile of chondrocytes in osteoarthritic knee using next-generation sequencing and bioinformatics. J Clin Med 7 (12):535
15. Castel D et al (2018) Small-RNA sequencing identifies dynamic microRNA deregulation during skeletal muscle lineage progression. Sci Rep 8(1):4208
16. Peffers MJ et al (2016) Decoding the regulatory landscape of ageing in musculoskeletal engineered tissues using genome-wide DNA methylation and RNASeq. PLoS One 11(8): e0160517
17. Outerbridge RE (1961) The etiology of chondromalacia patellae. J Bone Joint Surg Br 43-B:752–757
18. Chomczynski P, Sacchi N (1987) Single-step method of RNA isolation by acid guanidinium thiocyanate-phenol-chloroform extraction. Anal Biochem 162(1):156–159
19. Ro S et al (2006) A PCR-based method for detection and quantification of small RNAs. Biochem Biophys Res Commun 351 (3):756–763

Chapter 11

MicroRNA Expression Profiling, Target Identification, and Validation in Chondrocytes

Tomoki Chiba, Ryota Kurimoto, Takahide Matsushima, Yoshiaki Ito, Ryo Nakamichi, Martin Lotz, and Hiroshi Asahara

Abstract

MicroRNAs (miRNAs) are a class of noncoding small RNAs, which play a critical role in various biological processes including musculoskeletal formation and arthritis pathogenesis via regulating target gene expressions, raising the potentially substantial effects on gene expression networks. Over 2000 miRNAs are encoded in the human genome and a single miRNA potentially targets hundreds of genes. To examine the expression and function of miRNAs in chondrocytes and arthritis pathogenesis, we describe the protocols for the current miRNA related experiments including miRNA expression profiling by (1) Next Generation Sequencing and by TaqMan Array system, (2) miRNA target prediction by TargetScan, (3) miRNA target screening by cell-based reporter library assay, and (4) miRNA and its target interaction by HITS-CLIP (high-throughput sequencing of RNAs isolated by cross-linking immunoprecipitation) in cartilage and chondrocyte research.

Key words miRNA, Cartilage, Arthritis, Chondrocytes, Expression profiling, Target validation

1 Introduction

MicroRNAs (miRNAs) are a class of noncoding small RNAs that play a critical role in various biological processes, including musculoskeletal formation and arthritis pathogenesis, via regulating target gene expression [1–5]. During miRNA biogenesis, the primary transcripts (pri-miRNAs) transcribed by RNA polymerase II are processed by the nuclear ribonuclease Drosha in a DGCR8 complex in the nucleus, releasing ~60-bp-long hairpin precursor miRNAs (pre-miRNAs). Then, pre-miRNAs are processed by the ribonuclease Dicer, to ~22-nt long functionally mature miRNAs, which are incorporated into the RNA-induced silencing complex (RISC). The miRNA–RISC complex binds target mRNAs and mediates translational repression or degradation of mRNAs. This mRNA degradation and translational repression are mediated by sequence-specific interactions generally involving the 3′

Tariq M. Haqqi and Véronique Lefebvre (eds.), *Chondrocytes: Methods and Protocols*, Methods in Molecular Biology, vol. 2245, https://doi.org/10.1007/978-1-0716-1119-7_11,

untranslated region (UTRs) of the mRNA targets [6]. A single miRNA potentially targets hundreds of genes and may thereby have substantial effects on gene expression networks [7]. With over 2000 miRNAs encoded in the human genome, expression profiles and target identification are required for elucidating the role of each individual miRNA in health and disease [8]. Here, we describe a multistep protocol for miRNA expression profiling and validation including (1) Next Generation Sequencing and TaqMan Array system, (2) miRNA target prediction by TargetScan, (3) miRNA target screening by cell-based reporter library assay, and (4) miRNA/target interaction by HITS-CLIP (high-throughput sequencing of RNAs isolated by cross-linking immunoprecipitation) in cartilage and chondrocyte research.

2 Materials

2.1 Equipment

1. 7900HT Fast Real-Time PCR System (Thermo Fisher Scientific)
2. Next generation sequencer (e.g., MiSeq, Illumina).
3. Stratalinker 1800 (Stratagene).
4. Bioruptor UCD-250 (Sonic Bio).
5. Blue LED transilluminator (Thermo Fisher Scientific).
6. Bioanalyzer or TapeStation (Agilent Technologies).
7. Multidispenser (e.g., ECO DROPPER III; AS ONE).
8. Luminescence plate reader (e.g., ARVO X3; PerkinElmer).
9. ThermoMixer (Eppendorf).
10. Thermal cycler (Thermo Fisher Scientific).
11. Benchtop Centrifuge (Eppendorf).
12. Amersham Typhoon Biomolecular Imager (GE Healthcare).

2.2 Reagents and Kits

2.2.1 miRNA Sequence by NGS

1. TRIzol reagent (Thermo Fisher Scientific) or (TRI Reagent (Molecular Research Center) or ISOGEN (Nippongene).
2. TruSeq Small RNA Library Prep Kit (Illumina).
3. SYBR Safe DNA Gel Stain (Thermo Fisher Scientific).
4. GlycoBlue™ Coprecipitant (Thermo Fisher Scientific).
5. Novex™ TBE Gels, 10% (Thermo Fisher Scientific).
6. TBE Buffer (Tris–borate–EDTA) (10X) (Thermo Fisher Scientific).
7. Sodium Acetate (3 M), pH 5.5, RNase-free (Thermo Fisher Scientific).

2.2.2 miRNA Profiling by TaqMan Array System

1. Megaplex RT Primers (Thermo Fisher Scientific).
2. Megaplex PreAmp Primers (Thermo Fisher Scientific).
3. TaqMan MicroRNA array card (Thermo Fisher Scientific; These primers and array card are a set of 380 microRNAs and controls. There are A and B set in each species. It is necessary to select primers and arrays with the same number (A/B) and species.)
4. mirVana miRNA Isolation Kit (Thermo Fisher Scientific),
5. TRIzol reagent (Thermo Fisher Scientific) or (TRI Reagent (Molecular Research Center) or ISOGEN (Nippongene).
6. TaqMan® MicroRNA Reverse Transcription Kit (Thermo Fisher Scientific).
7. TaqMan® Universal PCR Master Mix, No AmpErase UNG (Thermo Fisher Scientific).

2.2.3 miRNA Target screening by Cell-Based Reporter Library Assay

1. Reporter library plates (*see* **Note 1**).
2. Pri-miRNA expression vector and control empty vector (*see* **Note 2**).
3. pRL-SV40 Renilla luciferase plasmid (Promega).
4. Elution buffer (EB): 10 mM Tris–HCl, pH 8.5.
5. FuGENE HD (Promega) (*see* **Note 3**).
6. Opti-MEM (Thermo Fisher Scientific).
7. Growth medium: DMEM (SIGMA) containing 10% fetal bovine serum (FBS) and 1% penicillin–streptomycin.
8. Wortmannin stock solution (*see* **Note 4**): 1 mM wortmannin (Sigma) in DMSO.
9. Dual-Glo Luciferase Assay System (Promega).
10. Phosphate buffered saline (PBS): 137 mM NaCl, 8.1 mM Na_2HPO_4, 2.68 mM KCl, 1.47 mM KH_2PO_4.
11. 293FT cells (Thermo Fisher Scientific).

2.2.4 miRNA and Target Interaction by CLIP

1. Lysis buffer; 1× PBS, 0.1% SDS, 0.5% deoxycholate, 0.5% NP-40, Complete™ Protease Inhibitor Cocktail.
2. Wash buffer; 1× PBS, 0.1% SDS, 0.5% deoxycholate, 0.5% NP-40.
3. High-salt wash buffer; 5× PBS, 0.1% SDS, 0.5% deoxycholate, 0.5% NP-40.
4. PNK buffer; 50 mM Tris–HCl (pH 7.5), 10 mM $MgCl_2$, 0.5% NP-40).
5. PNK buffer + EGTA; 50 mM Tris–HCl (pH 7.5), 20 mM EGTA, 0.5% NP-40.

6. Proteinase K solution; 4 mg/ml Proteinase K, 100 mM Tris–HCl (pH 7.5), 50 mM NaCl, 10 mM EDTA.
7. Urea PK buffer; 7 M urea, 100 mM Tris–HCl (pH 7.5), 50 mM NaCl, 10 mM EDTA.
8. 5× PNK Phosphatase Buffer (pH 6.5); 350 mM Tris–HCl (pH 6.5), 10 mM $MgCl_2$, 0.5% NP-40
9. Complete™ Protease Inhibitor Cocktail, EDTA-free (Roche).
10. 10 mM ATP (New England Biolabs)
11. NuPAGE™ LDS Sample Buffer (4×) (Thermo Fisher Scientific).
12. RNase A (Thermo Fisher Scientific).
13. RQ1 DNase (Promega).
14. Anti-pan Ago antibody (Millipore, clone 2A8).
15. Dynabeads Protein G (Thermo Fisher Scientific).
16. NuPAGE™ 10% Bis-Tris gel (*see* **Note 5**).
17. NuPAGE™ MOPS SDS Running Buffer (20×) (Thermo Fisher Scientific).
18. NuPAGE™ Transfer Buffer (20×) (Thermo Fisher Scientific).
19. Acid-phenol–chloroform, pH 4.5 (with IAA, 125:24:1) (Ambion).
20. Sodium Acetate (3 M), pH 5.5, RNase-free (Thermo Fisher Scientific).
21. RNasin® Ribonuclease Inhibitors (Promega).
22. Gel Loading Dye, Purple (6×) (New England Biolabs).
23. [γ-^{32}P] ATP (PerkinElmer)
24. T4 Polynucleotide Kinase (New England Biolabs).
25. T4 RNA ligase (New England Biolabs).
26. Proteinase K (Roche).
27. 3′ RNA linker (5′P-GUGUCAGUCACUUCCAGCGG-3′-amino acid)
28. 5′RNA linker (5′-OHAGGGAGGACGAUGCGG-3′OH)
29. SuperScript III (Thermo Fisher Scientific).
30. Primer DP5: 5′-AGGGAGGACGATGCGG-3′.
31. RT primer DP3: 5′-CCGCTGGAAGTGACTGACAC-3′.
32. Primer DSFP5: 5′-AATGATACGGCGACCACCGACTA TGGATACTTAGTCAGGGAGGACGATGCGG- 3′.
33. Primer DSFP3: 5′-CAAGCAGAAGACGGCATACGACCGC TGGAAGTGACTGACAC- 3′.
34. Phusion High-Fidelity DNA Polymerase (New England Biolabs).

35. KAPA Library Quantification Kit for Illumina (TakaRa).
36. QIAquick DNA Gel Extraction Kit (Qiagen).
37. UltraPure™ Low Melting Point Agarose (Thermo Fisher Scientific).

3 Methods

3.1 miRNA Expression Profiling

Several hundreds of miRNAs are expressed in cells and tissues, and their expressions are specific for each tissue and cell type. Understanding their pattern of expression is the first step in elucidating the function of miRNAs. Here, we will describe miRNA expression profiling by Next-Generation Sequencing and TaqMan Array. Expression profiling using Next Generation Sequencing can detect all miRNAs expressed in cells and tissues, and may also discover novel small RNAs [9–12]. On the other hand, TaqMan Array is a more sensitive, quantitative PCR-based method that can detect over a hundred of expressed miRNAs simultaneously. Thus, TaqMan Array might be useful for samples with limited amount of RNA such as clinical samples [13–15]. Several studies successfully demonstrated microRNA profiling by TaqMan Array in cell lines and tissues, including human cartilage [16–18].

3.1.1 miRNA Sequence by NGS

1. Total RNAs are purified using TRIzol reagent according to manufacturer's instructions (*see* **Note 6**).
2. To ligate the 3′ adaptor, mix 1 μl RA3 adaptor and 5 μl 200 ng/μl Total RNA in a PCR tube and incubate at 70 °C for 2 min.
3. Add 2 μl HML (ligation buffer), 1 μl RNase Inhibitor, and 1 μl T4 RNA Ligase 2. Incubate at 28 °C for 1 h.
4. Add 1 μl STP (stop solution) and incubate at 28 °C for 15 min.
5. To denature the 5′ adaptor, heat RA5 at 70 °C for 2 min before use.
6. Add 1 μl denatured RA5 adaptor, 1 μl ATP, and 1 μl T4 RNA Ligase and incubate at 28 °C for 1 h.
7. Transfer 6 μl adaptor-ligated RNA to a new PCR tube.
8. Add 1 μl RNA RT primer and incubate at 70 °C for 2 min.
9. Add 2 μl 5× First Strand Buffer, 0.5 μl 12.5 mM dNTP Mix, 1 μl 100 mM DTT, 1 μl RNase Inhibitor, 1 μl SuperScript II Reverse Transcriptase and incubate at 50 °C for 1 h.
10. Add 8.5 μl Ultrapure water, 5 μl PML (PCR mix), 2 μl RP1 (RNA PCR Primer), and 2 μl RNA PCR Primer Index.
11. Amplify library using the following program in thermal cycler: 98 °C for 30 s; 11 cycles of 98 °C for 10 s, 60 °C for 30 s, and 72 °C for 15 s; and 72 °C for 10 min.

12. Add DNA Loading Dye to amplified cDNA samples.
13. Load amplified cDNA samples onto a TBE PAGE gel.
14. Run the gel at 150 V for 1 h in 1xTBE buffer until blue dye reaches gel end.
15. Stain gel with SYBR Safe in 1xTBE at room temperature for 5 min.
16. View the gel on a blue LED transilluminator and cut out bands around 150 bp using a razor blade.
17. Crush the gel bands with pestle into small pieces in 1.5-ml tube and add 200 μl water.
18. Incubate at least 2 h with rotation.
19. Transfer the eluate and gel debris to the top of a 5-μm filter and centrifuge.
20. Add 4 μl GlycoBlue, 20 μl 3 M NaOAc, and 500 μl 100% ethanol.
21. Centrifuge at 20,000 × *g* for 20 min at 4 °C.
22. Discard the supernatant.
23. Wash the pellet with 500 μl 70% ethanol and centrifuge at 20,000 × *g* for 2 min.
24. Discard the supernatant and dry pellet briefly.
25. Resuspend the pellet in 10 μl 10 mM Tris–HCl, pH 8.5.
26. Check the library quality and quantity using a Bioanalyzer.
27. Dilute library to 2 nM with 10 mM Tris–HCl, pH 8.5.
28. Sequence at 15 pM final concentration on MiSeq.

3.1.2 miRNA Profiling by TaqMan Array System

1. Small RNAs are isolated using mirVana miRNA Kit. If needed, total RNA can be isolated using TRIzol reagent.
2. Total RNAs (350–1000 ng) are reverse-transcribed using Megaplex RT Primers and TaqMan® microRNA Reverse Transcription Kit. In brief, mix 0.8 μl Megaplex RT primer, 0.2 μl 100 mM dNTPs with dTTP 1.5 μl MultiScribe Reverse transcriptase, 0.8 μl 10xRT buffer, 0.9 μl 25 mM $MgCl_2$, 0.1 μl RNase Inhibitor, 1 μl total RNAs (350–1000 ng), and 2.7 μl nuclease-free water in a 1.5-ml microcentrifuge tube (*see* **Note 7**).
3. To synthesize cDNA, incubate as follows: 16 °C for 2 min; 40 cycles of 42 °C for 1 min and 50 °C for 1 s; and 85 °C for 5 min.
4. Mix 6 μl RT pTaqMan® Universal PCR Master Mix, No AmpErase UNG, and 444 μl nuclease-free water.

5. Load 100 μl mixture prepared in **step 4** into each port/reservoir of TaqMan array cards (2 loading ports to cover the total of 96 reaction chambers). Do not make air bubbles.
6. Centrifuge the card with 1200 rpm 1 min twice.
7. Seal the card.
8. Run a quantitative PCR on a 7900HT Fast Real-Time PCR System with TaqMan® Array Block using universal cycling conditions: 95 °C for 10 min; and then 40 cycles of 95 °C for 15 s and 60 °C for 60 s.
9. Raw Ct values are normalized with RNU6B, RNU44, or RNU48 Ct value. The resulting ΔCt values will then be used for subsequent analysis.

See also the manufacturer's protocol below.

1. http://tools.thermofisher.com/content/sfs/manuals/cms_042326.pdf
2. https://tools.thermofisher.com/content/sfs/manuals/4399721c.pdf

3.2 miRNA Target Prediction by TargetScan

To investigate a predetermined gene as a target of miRNA regulation, use computed bioinformatic algorithms that predict the probability of a functional miRNA binding site. For example, miRbase (http://www.mirbase.org/), microRNA.org (http://www.microrna.org/microrna/home.do), TargetScan (http://www.targetscan.org/vert_72/), and miRWalk (http://mirwalk.umm.uni-heidelberg.de/) [19–25]. Most programs use 3′-UTR datasets to look for target sites but programs like miRWalk include the 5′-UTR, coding (CDS), and 3′-UTR sequences. These algorithms often differ from one another in their miRNA binding site predictions. Thus, it is recommended to use multiple algorithms and to validate the findings. In this paragraph, we describe a basic approach using TargetScan.

1. Open http://www.targetscan.org/vert_71/ (Fig. 1-1).
2. Select a desired mRNA species as target in "2" (Fig. 1-2).
3. Enter a miRNA name in "3" (Fig. 1-3).
4. Click "submit" (Fig. 1-4).
5. Click "Download table" to get the data in a "txt" or "xlsx" file (Fig. 2).

3.3 miRNA Target screening Using a Cell-Based Reporter Library Assay

Target identification is necessary for understanding miRNA functions. A combination of transcriptome analysis and in silico prediction tools is often used to identify targets. A transcriptome analysis, performed for instance using an RNA-seq assay, for samples generated in the presence or absence of a specific miRNA, can help find

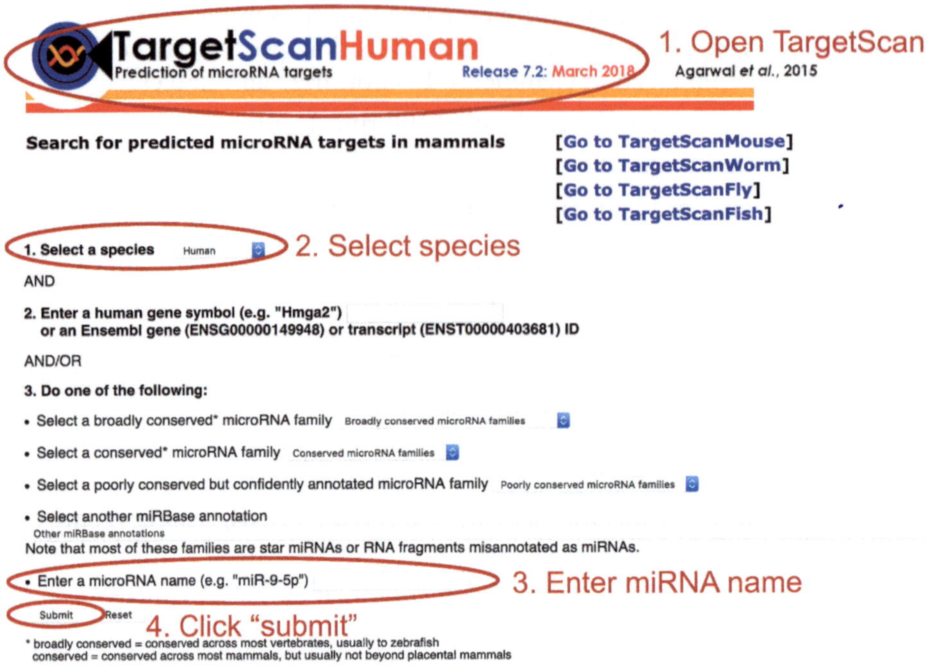

Fig. 1 Start page of TargetScan

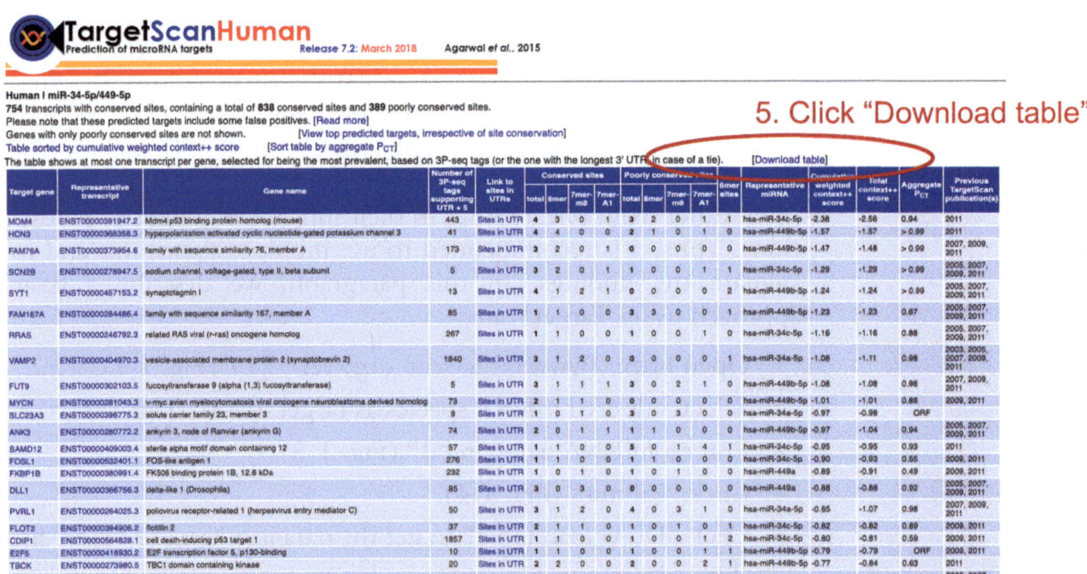

Fig. 2 Next page of TargetScan; predicted miRNA binding sites are listed

putative target mRNAs, provided that these RNAs have their stability significantly decreased by the miRNA. In contrast, miRNA targets regulated at the translational level cannot always be identified based on protein levels [26–28]. Current computational tools for prediction of miRNA targets, such as TargetScan [29], predict targets by miRNA–mRNA sequence matches. However, the identified candidate targets often contain many false positives, and are often identified solely based on analysis of the 3′-UTR of mRNAs. On the other hand, a novel approach for identifying miRNA targets has been reported that consist in a cell-based screening system. This system uses reporter vector libraries with or without the specific miRNAs. It is thus a functional assay to detect miRNA targets in a high-throughput manner and can evaluate putative miRNA targets not only at the mRNA level but also at the protein level. 3′LIFE, reported by Wolter et al., is a cell-based miRNA target screening system based on a luciferase reporter library of 275 human 3′-UTRs. This system was used to rapidly and sensitively identify the targets of let-7c and miR-10b [30, 31]. In addition, the same group constructed a luciferase reporter library of 1461 human 3′-UTRs, termed "human 3′UTRome v1 clone collection" (h3′UTRome v1), which consists of human 3′-UTRs from transcription factors, kinases, and RNA-binding proteins [32]. This system allows for screening individual miRNAs without biasing the screen toward candidate genes identified bioinformatically. However, the 3′-UTR library is not sufficiently large-scale and the target regions are restricted to the 3′-UTRs. To overcome this problem, our group developed a reporter library system. We constructed a reporter plasmid library in which the luciferase gene includes 4891 nonbiased full-length cDNA sequences in its 3′-UTR [33]. This system allowed us to evaluate the putative direct targets of specific miRNAs functionally through its full-length sequence at the protein level. Using this system, the targets of miR-34a, a tumor suppressor miRNA, were screened and both reported and novel miR-34a targets were identified [33]. Importantly, the expressions of several identified targets were not reduced in miR-34a-overexpressing cells at the mRNA levels but were reduced at the protein levels, indicating that this system can identify the targets of the miRNA that are regulated at the translation level [33]. Furthermore, several identified targets were directly regulated by miR-34a via their coding region, highlighting the significance of reporters containing full-length cDNA sequences [33]. Here, we describe the detailed procedure of the cell-based miRNA target screening.

1. Mix 30 ml of Opti-MEM and 0.9 ml of FuGENE HD in a 50-ml tube (two tubes) and incubate for 5 min.
2. Add 30 μg of pRL-SV40 plasmid and 120 μg of pri-miRNA expression vector or control empty vector.
3. Add 5 μl of the mixture to reporter library plates into each well using a multi-dispenser and incubate for 20 min.

4. Add 5000 293FT cells in 40 μl of growth medium with a multidispenser in each well and culture the cells in a 5% CO_2 incubator at 37 °C for 24 h.
5. Mix 50 ml of growth medium and 5 μl of wortmannin stock solution and add 5 μl of the mixture into each well (final concentration of 100 nM) and continue the culture in the 5% CO_2 incubator at 37 °C for 5 h.
6. Mix 50 ml of Dual-Glo buffer and substrate mixture with 50 ml of PBS.
7. Discard the medium of culture plates and add 10 μl of Dual-Glo–PBS mixture in each well and then incubate for 10 min at room temperature.
8. Measure the *firefly* luciferase activities with a luminescence plate reader (*see* **Note 8**).
9. Mix 50 ml of Stop&Glo buffer with 0.5 ml of Stop&Glo substrate.
10. Add 5 μl of Stop&Glo mixture in each well and then incubate for 10 min at room temperature.
11. Measure the *Renilla* luciferase activities with a luminescence plate reader.
12. Normalize the *firefly* luciferase activities with the *Renilla* luciferase activities. Select target candidates whose reporter activity is reduced by a pri-miRNA expression vector compared to the empty control (*see* **Note 9**).

3.4 miRNA and Target Interaction by CLIP

HITS-CLIP (high-throughput sequencing of RNAs isolated by cross-linking immunoprecipitation) combines biochemical methods with high-throughput sequencing technology to identify RNA–protein interactions [34]. When applied with an antibody against the RNA-induced silencing complex (RISC) factor Argonaute (AGO), this method can detect miRNA-targeted mRNAs through the isolation of endogenous AGO–miRNA–mRNA complexes [35–37]. Here, we will describe a basic method of Ago-CLIP with cultured cells.

3.4.1 Preparation of Antibody-Coupled Beads

1. Dispense 50 μl of Dynabeads Protein G per sample in a 1.5-ml tube and wash the beads with Lysis buffer 3 times.
2. Suspend beads in 100 μl of Lysis buffer, add 5 μg of anti-pan-Ago antibody per sample, and incubate at room temperature for 1 h with rotation.
3. Wash beads with Lysis buffer 3 times.
4. Suspend beads in 50 μl Lysis buffer.

3.4.2 HITS-CLIP

UV Cross-Linking and Immunoprecipitation

1. Grow cells in a 100- or 150-mm plate and rinse once with PBS.
2. Place in a Stratalinker instrument with the cover off and irradiate with 254 nm UV at 200 mJ/cm^2 or 400 mJ/cm^2 for RNA–protein cross-linking.
3. Collect the cells with a cell scraper and centrifuge at 2500 x g for 5 min.
4. Suspend the cells in 1 ml Lysis buffer and incubate at 4 °C for 10 min with rotation.
5. Sonicate the cell lysate with a Bioruptor using 30 s-on/30 s-off procedure for 5 cycles at "low" setting.
6. Add 5 μl diluted RNase A and 2 μl of RQ1 DNase, incubate for 5 min at 37 °C, and then keep on ice. We recommend preparing two different conditions of RNase A, typically 1:100 and 1:10,000. In addition, non–cross-linked cells are used as negative controls.
7. Centrifuge at 13,000 × *g* for 15 min at 4 °C and transfer the supernatant to new 1.5-ml tube.
8. Add 50 μl antibody-coupled beads and incubate at 4 °C for 2 h or overnight with rotation.
9. Wash beads with Wash buffer and High-salt wash buffer.
10. Suspend beads in PNK Phosphatase mixture as follows: 15 μl nuclease-free water, 4 μl 5xPNK buffer pH 6.5, and 1 μl T4 PNK.
11. Incubate at 37 °C for 20 min in a ThermoMixer at 1100 rpm.
12. Wash beads with PNK buffer, PNK + EGTA buffer, and PNK buffer twice.
13. Suspend beads in 3′ linker ligation mixture as follows: 10.9 μl nuclease-free water, 3 μl 10xT4 RNA ligase buffer, 3 μl 10 mM ATP, 0.3 μl 20 mg/ml BSA, 0.8 μl 100% DMSO, 9 μl 50% PEG8000, 1 μl 100 μM 3′ linker, and 2 μl T4 RNA ligase.
14. Incubate in a ThermoMixer at 16 °C overnight.
15. Wash beads sequentially with PNK buffer, High-salt wash buffer, and PNK buffer.

End Labeling and Reverse Transcription

1. Suspend beads in PNK labeling mixture as follows: 32 μl nuclease-free water, 4 μl 10xPNK buffer, 2 μl T4 PNK, and 2 μl [γ-^{32}P] ATP.
2. Incubate in ThermoMixer at 37 °C for 20 min.
3. Add 2 μl 10 mM ATP and incubate in ThermoMixer at 37 °C for another 5 min.
4. Wash beads sequentially with PNK buffer, High-salt wash buffer, and PNK buffer.

5. Suspend beads in 1× NuPAGE LDS Sample Buffer and incubate at 70 °C for 10 min.
6. Let stand on a magnetic stand and load the supernatant on NuPAGE 10% gel.
7. Run at 150 V for 60 min.
8. Transfer gel to nitrocellulose membrane in 1x NuPAGE Transfer Buffer with 10% methanol at 30 V for 1 h in the cold room.
9. Expose the membrane to an imager plate in a cassette for 1 h.
10. Visualize the autoradiogram with a bioimaging analyzer. We can usually see two different modal sizes, one at ~110 kDa corresponding to Ago–miRNA complexes and the other at ~130 kDa corresponding to Ago–mRNA complexes.
11. Cut the bands on the membrane and put them in a 1.5-ml tube.
12. Add 200 μl proteinase K solution and incubate in a ThermoMixer at 37 °C for 20 min.
13. Add 200 μl 7 M Urea in PK Buffer and incubate in a ThermoMixer at 37 °C for 20 min.
14. Add 400 μl acid-phenol–chloroform, mix using a vortex mixer, and incubate in a ThermoMixer at 37 °C for 20 min.
15. Centrifuge at 20,000 × *g* for 5 min and collect the aqueous phase.
16. Add 50 μl 3 M NaOAc (pH 5.5), 1 ml 1:1 mix of ethanol and isopropanol, and 1.5 μl of GlycoBlue.
17. Precipitate at −20 °C overnight.
18. Centrifuge at 20,000 × *g* for 15 min.
19. Wash with 70% ethanol.
20. Dissolve in 10.9 μl nuclease-free water.
21. Add 3 μl 10X T4 RNA ligase buffer, 0.3 μl 20 mg/ml BSA, 3 μl 10 mM ATP, 0.8 μl DMSO, 9 μl 50% PEG8000, 2 μl T4 RNA ligase, and 1 μl 2 μM 5′ RNA linker.
22. Incubate at 16 °C overnight.
23. Add 370 μl nuclease-free water and 400 μl acid-phenol–chloroform, mix with a vortex mixer and incubate in a ThermoMixer at 37 °C for 20 min.
24. Centrifuge at 20,000 × *g* for 5 min and collect the aqueous phase.
25. Add 50 μl 3 M NaOAc (pH 5.5), 1 ml of 1:1 mix of ethanol and isopropanol, and 1.5 μl of GlycoBlue.
26. Precipitate at −30 °C overnight.
27. Centrifuge at 20,000 × *g* for 15 min.

28. Wash with 70% ethanol.
29. Dissolve in 8 μl nuclease-free water.
30. Add 2 μl 5 nM DP3 and 3 μl 3 mM dNTPs and incubate at 65 °C for 5 min.
31. Add 1 μl 0.1 M DTT, 4 μl 5X SuperScript RT Buffer, 1 μl RNAsin, and 1 μl SuperScript III.
32. Incubate at 50 °C for 45 min, at 55 °C for 15 min, and then at 90 °C for 5 min.

Library Enrichment and Sequencing

1. Mix as follows: 5.5 μl nuclease-free water, 12.5 μl 2× PhusionHF Buffer, 0.5 μl 20 nM DP3, 0.5 μl 20 nM DP5, 5 μl cDNA, and 1 μl Phusion DNA polymerase.
2. Amplify the library using the following program in a thermal cycler: 98 °C for 30 s; 20 to 25 cycles of 98 °C for 10 s, 60 °C for 10 s, and 72 °C for 10 s; and 72 °C for 10 min.
3. Add DNA gel loading dye to the PCR product.
4. Run a 2% agarose gel at 100 V for 30 min.
5. Stain the gel with SYBR Gold in 1× TBE.
6. Cut out the band containing 80–100 nt-long DNA, and extract the DNA with QIAquick Gel Extraction Kit.
7. Mix as follows: 5.5 μl nuclease-free water, 12.5 μl 2× PhusionHF Buffer, 0.5 μl 20 nM DSFP5, 0.5 μl 10 nM DSFP3, 5 μl first PCR product, and 1 μl Phusion DNA polymerase.
8. Amplify the library using the following program in a thermal cycler: 98 °C for 30 s; 20 to 25 cycles of 98 °C for 10 s, 60 °C for 10 s, and 72 °C for 10 s; and 72 °C for 10 min.
9. Add DNA gel loading dye to the PCR product.
10. Run a 2% agarose gel at 100 V for 30 min.
11. Stain the gel with SYBR Gold in 1× TBE.
12. Cut out the DNA band at 150–200 nt and extract the DNA with QIAquick Gel Extraction Kit.
13. Check library quality in a Bioanalyzer.
14. Quantify the library using Kapa Library Quantification Kit.
15. Dilute the Ago-CLIP library to 2 nM with 10 mM Tris–HCl, pH 8.5.
16. Sequence at a final concentration of 15 pM on MiSeq.

4 Notes

1. A reporter library in which the luciferase gene includes 4891 nonbiased full-length cDNA sequences in its 3′-UTR [33]. Purify the reporter library plasmids by general column cleanup kit such as LaboPass Mini (Hokkaido System Science). Dilute the reporter plasmid library with EB in 96-well plates. Then, dispense the plasmids into 384-well culture plates using a multidispenser (e.g., EDR-384SII [BioTec]).
2. Clone the desired pri-miRNA sequence into a mammalian expression vector (e.g., pcDNA3.1 (Invitrogen)). Check the expression of the mature miRNA from the vector by TaqMan quantitative PCR or northern blotting. Also, check the plasmid in an assay with a reporter containing an already identified target.
3. Polyethylenimine MAX can also be used as the transfection reagent.
4. Wortmannin is an inhibitor of nonsense-mediated mRNA decay (NMD) [38, 39]. The function of NMD is to reduce errors in gene expression by eliminating mRNAs that contain premature termination codons (PTCs) [40, 41]. In addition, the length of the 3′-UTR influences the NMD pathway. A number of studies have revealed that artificial 3′-UTRs of 800–900 bp in length promote NMD [42–49]. Since the average length of their 3′-UTRs is more than 2000 bp, our reporters may be affected by NMD. We found that the repression of reporter activity by a miRNA expression plasmid is more clearly detected when the NMD inhibitor wortmannin is added (8).
5. The pH of standard SDS-PAGE gels (Laemmli) changes during the run and can reach up to 9.5. In contrast, the Novex NuPAGE buffer system keeps the pH around 7 throughout the run. We use Bis-Tris gels from this system, and a MOPS buffer to avoid alkaline hydrolysis of the RNA.
6. To isolate RNAs, a silica membrane-based purification column can be used in addition to the TRIzol reagent. Since columns designed for total RNA purification lose RNAs shorter than 300 nucleotides, it is necessary to use a purification column specifically designed for small RNAs, such as mirVana (Thermo Fisher Scientific). In addition, small RNAs purified using such a column can also be used as input for next-generation sequencing and TaqMan Array assays.
7. If the total amount of RNA is less than 350 ng, a preamplification step with Megaplex PreAmp Primers and TaqMan PreAmp Master Mix, 2×, is necessary after the RT reaction.

8. Before screening, luciferase assay conditions need to be determined. We usually perform the screening with 10–30 ng of reporter plasmids, 20–50 ng of effector plasmids, 2.5–5 ng of pRL-SV40 plasmid, and 0.1–0.2 μl of transfection reagent in each well, with a final concentration of wortmannin of 100 nM.
9. Target candidate reporters should be confirmed in luciferase assays at a 96-well plate scale.

References

1. Miyaki S, Asahara H (2012) Macro view of microRNA function in osteoarthritis. Nat Rev Rheumatol 8:543–552
2. Inui M, Mokuda S, Sato T et al (2018) Dissecting the roles of miR-140 and its host gene. Nat Cell Biol 20:516–518
3. Nakasa T, Miyaki S, Okubo A et al (2008) Expression of microRNA-146 in rheumatoid arthritis synovial tissue. Arthritis Rheum 58:1284–1292
4. Miyaki S, Sato T, Inoue A et al (2010) MicroRNA-140 plays dual roles in both cartilage development and homeostasis. Genes Dev 24:1173–1185
5. Miyaki S, Nakasa T, Otsuki S et al (2009) MicroRNA-140 is expressed in differentiated human articular chondrocytes and modulates interleukin-1 responses. Arthritis Rheum 60:2723–2730
6. Bartel DP (2018) Metazoan MicroRNAs. Cell 173:20–51
7. Lim LP, Lau NC, Garrett-Engele P et al (2005) Microarray analysis shows that some microRNAs downregulate large numbers of target mRNAs. Nature 433:769–773
8. Kozomara A, Birgaoanu M, Griffiths-Jones S (2019) miRBase: from microRNA sequences to function. Nucleic Acids Res 47:D155–D162
9. Creighton CJ, Reid JG, Gunaratne PH (2009) Expression profiling of microRNAs by deep sequencing. Brief Bioinform 10:490–497
10. Morin RD, O'Connor MD, Griffith M et al (2008) Application of massively parallel sequencing to microRNA profiling and discovery in human embryonic stem cells. Genome Res 18:610–621
11. Chiang HR, Schoenfeld LW, Ruby JG et al (2010) Mammalian microRNAs: experimental evaluation of novel and previously annotated genes. Genes Dev 24:992–1009
12. Hu Y, Lan W, Miller D (2017) Next-generation sequencing for MicroRNA expression profile. Methods Mol Biol 1617:169–177
13. Chen C, Ridzon DA, Broomer AJ et al (2005) Real-time quantification of microRNAs by stem-loop RT-PCR. Nucleic Acids Res 33: e179
14. Mestdagh P, Feys T, Bernard N et al (2008) High-throughput stem-loop RT-qPCR miRNA expression profiling using minute amounts of input RNA. Nucleic Acids Res 36: e143
15. Grigorenko EV, Ortenberg E, Hurley J et al (2011) miRNA profiling on high-throughput OpenArrayTM system. In: Wu W (ed) MicroRNA and cancer: methods and protocols. Humana Press, Totowa, NJ, pp 101–110
16. Chen Y, Gelfond JAL, McManus LM, Shireman PK (2009) Reproducibility of quantitative RT-PCR array in miRNA expression profiling and comparison with microarray analysis. BMC Genomics 10:407
17. Hui ABY, Shi W, Boutros PC et al (2009) Robust global micro-RNA profiling with formalin-fixed paraffin-embedded breast cancer tissues. Lab Investig 89:597–606
18. McAlinden A, Varghese N, Wirthlin L, Chang L-W (2013) Differentially expressed microRNAs in chondrocytes from distinct regions of developing human cartilage. PLoS One 8: e75012
19. Ambros V, Bartel B, Bartel DP et al (2003) A uniform system for microRNA annotation. RNA 9:277–279
20. Griffiths-Jones S (2004) The microRNA Registry. Nucleic Acids Res 32:D109–D111
21. Enright AJ, John B, Gaul U et al (2003) MicroRNA targets in drosophila. Genome Biol 5:R1
22. John B, Enright AJ, Aravin A et al (2004) Human microRNA targets. PLoS Biol 2:e363
23. Lewis BP, Burge CB, Bartel DP (2005) Conserved seed pairing, often flanked by adenosines, indicates that thousands of human genes are microRNA targets. Cell 120:15–20
24. Grimson A, Farh KK-H, Johnston WK et al (2007) MicroRNA targeting specificity in

mammals: determinants beyond seed pairing. Mol Cell 27:91–105

25. Sticht C, De La Torre C, Parveen A, Gretz N (2018) miRWalk: an online resource for prediction of microRNA binding sites. PLoS One 13:e0206239
26. Bhattacharyya SN, Habermacher R, Martine U et al (2006) Relief of microRNA-mediated translational repression in human cells subjected to stress. Cell 125:1111–1124
27. Bartel DP (2009) MicroRNAs: target recognition and regulatory functions. Cell 136:215–233
28. Meijer HA, Kong YW, Lu WT et al (2013) Translational repression and eIF4A2 activity are critical for microRNA-mediated gene regulation. Science 340:82–85
29. Lewis BP, Shih I-H, Jones-Rhoades MW et al (2003) Prediction of mammalian microRNA targets. Cell 115:787–798
30. Wolter JM, Kotagama K, Pierre-Bez AC et al (2014) 3'LIFE: a functional assay to detect miRNA targets in high-throughput. Nucleic Acids Res 42:e132
31. Wolter JM, Kotagama K, Babb CS, Mangone M (2015) Detection of miRNA targets in high-throughput using the 3'LIFE assay. J Vis Exp (99):e52647
32. Kotagama K, Babb CS, Wolter JM et al (2015) A human 3'UTR clone collection to study post-transcriptional gene regulation. BMC Genomics 16:1036
33. Ito Y, Inoue A, Seers T et al (2017) Identification of targets of tumor suppressor microRNA-34a using a reporter library system. Proc Natl Acad Sci U S A 114:3927–3932
34. Licatalosi DD, Mele A, Fak JJ et al (2008) HITS-CLIP yields genome-wide insights into brain alternative RNA processing. Nature 456:464–469
35. Chi SW, Zang JB, Mele A, Darnell RB (2009) Argonaute HITS-CLIP decodes microRNA-mRNA interaction maps. Nature 460:479–486
36. Moore MJ, Zhang C, Gantman EC et al (2014) Mapping Argonaute and conventional RNA-binding protein interactions with RNA at single-nucleotide resolution using HITS-CLIP and CIMS analysis. Nat Protoc 9:263–293
37. Loeb GB, Khan AA, Canner D et al (2012) Transcriptome-wide miR-155 binding map reveals widespread noncanonical microRNA targeting. Mol Cell 48:760–770
38. Pal M, Ishigaki Y, Nagy E, Maquat LE (2001) Evidence that phosphorylation of human Upfl protein varies with intracellular location and is mediated by a wortmannin-sensitive and rapamycin-sensitive PI 3-kinase-related kinase signaling pathway. RNA 7:5–15
39. Usuki F, Yamashita A, Kashima I et al (2006) Specific inhibition of nonsense-mediated mRNA decay components, SMG-1 or Upfl, rescues the phenotype of Ullrich disease fibroblasts. Mol Ther 14:351–360
40. Frischmeyer PA, Dietz HC (1999) Nonsense-mediated mRNA decay in health and disease. Hum Mol Genet 8:1893–1900
41. Hug N, Longman D, Cáceres JF (2016) Mechanism and regulation of the nonsense-mediated decay pathway. Nucleic Acids Res 44:1483–1495
42. Toma KG, Rebbapragada I, Durand S, Lykke-Andersen J (2015) Identification of elements in human long 3' UTRs that inhibit nonsense-mediated decay. RNA 21:887–897
43. Eberle AB, Stalder L, Mathys H et al (2008) Posttranscriptional gene regulation by spatial rearrangement of the 3′ untranslated region. PLoS Biol 6:e92
44. Singh G, Rebbapragada I, Lykke-Andersen J (2008) A competition between stimulators and antagonists of Upf complex recruitment governs human nonsense-mediated mRNA decay. PLoS Biol 6:e111
45. Ruiz-Echevarría MJ, Peltz SW (2000) The RNA binding protein Pub1 modulates the stability of transcripts containing upstream open reading frames. Cell 101:741–751
46. Hogg JR, Goff SP (2010) Upfl senses 3'UTR length to potentiate mRNA decay. Cell 143:379–389
47. Huang L, Lou C-H, Chan W et al (2011) RNA homeostasis governed by cell type-specific and branched feedback loops acting on NMD. Mol Cell 43:950–961
48. Hurt JA, Robertson AD, Burge CB (2013) Global analyses of UPF1 binding and function reveal expanded scope of nonsense-mediated mRNA decay. Genome Res 23:1636–1650
49. Yepiskoposyan H, Aeschimann F, Nilsson D et al (2011) Autoregulation of the nonsense-mediated mRNA decay pathway in human cells. RNA 17:2108–2118

Chapter 12

ChIP-Seq Assays from Mammalian Cartilage and Chondrocytes

Akira Yamakawa, Hironori Hojo, and Shinsuke Ohba

Abstract

Chromatin immunoprecipitation followed by sequencing (ChIP-seq) is a powerful tool to identify binding profiles of transcriptional regulators and chromatin regulators as well as histone modification patterns in a genome-wide manner. ChIP-seq consists of five major steps: (1) preparation of cells and chromatin, (2) ChIP, (3) ChIP-seq library construction, (4) sequencing of ChIP DNA with a next-generation sequencer (NGS), and (5) computational analysis of sequence data. Recent ChIP-seq studies in skeletal tissues enable us to understand the modes of action of key skeletal regulators, functional interaction among the enhancers bound by the regulators, the complex nature of regulatory inputs, and thereby the gene regulatory landscape in skeletal development. Here we describe a ChIP-seq protocol that we have employed in our studies, with particular focus on chromatin preparation and subsequent ChIP in skeletal cells, including chondrocytes.

Key words Cartilage, Chondrocyte, Bone, Osteoblast, Skeleton, ChIP, ChIP-seq, Next-generation sequencer

1 Introduction

Cell identity and the coordination of developmental processes are governed by gene regulatory networks. Both enhancers and promoters play pivotal roles in the networks; enhancers are defined as DNA sequences that have potential to mediate the spatial and temporal pattern of gene expressions regardless of the orientation of the sequence, that is, forward or reverse, and of the location of the sequence. They can function at megabase distance from transcriptional start sites (TSSs). Recent comprehensive genome-wide studies have identified binding profiles of transcriptional regulators and chromatin regulators as well as histone modification patterns at enhancers and promoters. These findings provide us with insight into the gene regulatory networks across a large number of cell types in various species including flies, mice, and humans [1–5]. Integrated studies on these data are a key to understanding

Tariq M. Haqqi and Véronique Lefebvre (eds.), *Chondrocytes: Methods and Protocols*, Methods in Molecular Biology, vol. 2245, https://doi.org/10.1007/978-1-0716-1119-7_12,

the cell-fate specificity and dynamic differentiation of cell types of interest.

Key transcriptional regulators for skeletal development, including Sox9, Sox5/6, Runx2, AP-1, and Sp7/osterix, have been identified; their modes of action have been studied with particular focus on local gene regulatory networks. Recent evidences emerging from extensive analyses with chromatin immunoprecipitation followed by sequencing (ChIP-seq) provide novel insight into the modes of action of the key regulators, functional interaction among the enhancers bound by the regulators, and the complex nature of regulatory inputs in skeletal development, leading to rapid progress in our understanding of the gene regulatory landscape on the chondrocyte and osteoblast genome [6–15].

In this chapter, we describe a ChIP-seq protocol that we have employed in our studies, with particular focus on chromatin preparation and subsequent ChIP, rather than on sequencing procedures with the next-generation sequencer (NGS) and computational analysis of sequence data. We modified and optimized published protocols [16–18] to obtain high-quality ChIP-seq data efficiently from skeletal cells. The protocol has enabled us to identify binding profiles of transcriptional regulators as well as histone modification patterns in mouse primary chondrocytes and osteoblasts [11–13]. The protocol can also be applied to skeletal cell lines cultured in vitro [12].

2 Materials

Prepare all solutions using ultrapure water and analytical grade reagents, and store as indicated. Also note that some reagents need to be freshly prepared each time.

2.1 Equipment

1. 70 μm cell strainers
2. 15 mL and 50 mL polypropylene tubes
3. 1.5 mL tubes
4. Pipettes.
5. Scissors and forceps.
6. Centrifuge.
7. Microscope.
8. Thermal incubator.
9. Platform shaker.
10. Magnetic stand.
11. Rotating platform.
12. Digital Sonifier System (Branson).

13. Qubit™ Fluorometer System (Thermo Fisher Scientific).
14. Agilent 2100 Bioanalyzer (Agilent Technologies).

2.2 Reagents for Isolation of Primary Rib Chondrocytes from Newborn Mice and Cross-Link

1. Magnesium- and calcium-free phosphate buffered saline (PBS (-)).
2. 2 mg/mL Pronase (from *Streptomyces griseus*) in PBS (-): Prepare freshly each time.
3. 2 mg/mL collagenase D (from *Clostridium histolyticum*) in PBS (-): Prepare freshly each time.
4. Fetal bovine serum (FBS).
5. EDTA-free protease inhibitor cocktail (Merck).
6. 37% formaldehyde
7. 2.5 M glycine.

2.3 Reagents for ChIP

1. 0.5% BSA–PBS: Dissolve bovine serum albumin (BSA) fraction V in PBS (-). Prepare freshly each time and keep on ice.
2. Magnetic beads (Dynabeads®; Thermo Fisher Scientific).
3. Antibodies for proteins of interest.
4. EDTA-free protease inhibitor cocktail (Merck).
5. Lysis buffer: 50 mM HEPES–KOH (pH 7.5), 140 mM NaCl, 1 mM EDTA, 10% glycerol, 0.5% IGEPAL CA-630 (or NP-40), and 0.25% Triton X-100. Keep on ice. Add the protease inhibitor cocktail immediately before use.
6. Buffer 2: 10 mM Tris–HCl (pH 8.0), 200 mM NaCl, 1 mM EDTA, and 0.5 mM EGTA. Keep at room temperature. Add the protease inhibitor cocktail immediately before use.
7. Buffer 3: 10 mM Tris–HCl (pH 8.0), 1 mM EDTA, and 0.5 mM EGTA. Keep on ice. Add the protease inhibitor cocktail immediately before use.
8. 20% N-lauroylsarcosine sodium solution: Dissolve N-lauroylsarcosine sodium salt in Buffer 3. Prepare freshly each time and keep on ice.
9. ChIP cocktail mix: 433 mM NaCl, 0.43% sodium deoxycholate, and 4.3% Triton X-100. Prepare freshly each time and keep on ice.
10. RIPA buffer: 50 mM HEPES-KOH (pH 7.5), 500 mM LiCl, 1 mM EDTA, 0.7% sodium deoxycholate, and 1% IGEPAL CA-630 (or NP-40). Keep on ice. Add the protease inhibitor cocktail immediately before use.
11. 50 mM NaCl–TE (Tris–EDTA) buffer (TE: 10 mM Tris–HCl (pH 8.0), and 1 mM EDTA (pH 8.0)). Keep on ice. Add the protease inhibitor cocktail immediately before use.

12. Elution buffer: 50 mM Tris–HCl (pH 8.0), 10 mM EDTA, and 1% SDS. Prepare freshly each time.
13. RNase A.
14. Proteinase K.
15. MinElute PCR Purification kit (Qiagen) or equivalent column systems designed for small elution volumes.
16. Qubit™ dsDNA HS Assay Kit.
17. High Sensitivity DNA Kit for the Agilent 2100 Bioanalyzer.

2.4 Reagents for ChIP-Seq Library Construction

1. Library construction kits or equivalent sets of reagents and equipment that are designed for the NGS platforms used.
2. Agencourt AMPure XP.

3 Methods

3.1 Isolation of Primary Rib Chondrocytes from Newborn Mice and Cross-Link

Isolation of Primary Rib Chondrocytes from Newborn Mice

1. Sacrifice newborn mice under general anesthesia.
2. Place the mice in the back-down position on the table.
3. Make a longitudinal incision through the skin of the abdomen and rib cage using scissors and forceps.
4. Cut the diaphragm and eliminate all the soft tissues from the rib cage.
5. Dissect the ventral part of the ribs from the vertebrae using scissors.
6. Place the isolated ribs in a 50 mL tube containing cold PBS(-).
7. Remove the soft tissues using forceps and dissect the ribs using scissors under a microscope. Do not include calcified regions of the rib cartilage (*see* **Note 1**).
8. Rinse the ribs twice with cold PBS(-).
9. Incubate pieces of the rib cartilage in the Pronase solution for 60 min at 37 °C in a thermal incubator. Agitate tissue fragments every 30 min, using 25 mL or 10 mL pipettes, to detach soft tissues.
10. Make sure that all soft tissues are detached from the fragmented rib cartilage. If not, extend the incubation time.
11. Exchange the Pronase solution with PBS(-) and mix gently. After the rib cartilage has settled at the bottom of the tube, discard the supernatant. Repeat 5 times until the rib cartilage seems to be clean.
12. Resuspend the rib cartilage in the Collagenase D solution and incubate at 37 °C in a thermal incubator for 90 min or until the

cartilage is completely digested. Agitate tissue fragments every 30 min using 25 mL or 10 mL pipettes.

13. After adding FBS at a final concentration of 10%, filter the cell suspension with a 70 μm cell strainer and collect the flow-through in a 50 mL tube.
14. Centrifuge at 400 × *g* for 5 min and discard the supernatant (*see* **Note 2**).
15. Resuspend the cell pellet in PBS(-).
16. Count the cell number with a hemocytometer and make aliquots in 15 mL tubes, each of which contains 10^7 cells (*see* **Note 3**).
17. Centrifuge at 400 × *g* for 5 min and discard the supernatant.
18. Resuspend the cell pellet in 10 mL PBS(−).

Cross-link

19. Add 270 μL of the 37% formaldehyde solution to the suspension of cells in 10 mL PBS (-).
20. Cross-link for 10 min at room temperature with mild agitation on a platform shaker (*see* **Note 4**).
21. Add 500 μL of 2.5 M glycine and incubate for 5 min at room temperature with mild agitation on a platform shaker.
22. Centrifuge at 400 × *g* for 5 min at 4 °C and discard the supernatant.
23. Resuspend the cell pellet in 5 mL cold PBS (−) supplemented with the protease inhibitor cocktail.
24. Centrifuge at 400 × *g* for 5 min at 4 °C and discard the supernatant.
25. Repeat **steps 23** and **24**.
26. Directly proceed to Subheading 3.2, **step 9** or store the cell pellets at −80 °C in 15 mL tubes until ChIP.

3.2 ChIP

Day 1

Preparation of Antibody-Conjugated Magnetic Beads

1. Transfer 100 μL Dynabeads® into 1 mL of 0.5% BSA–PBS in a 1.5 mL tube.
2. Place the tube on the magnetic stand to collect beads for 2–3 min and aspirate the liquid.
3. Resuspend the beads in 1 mL of 0.5% BSA–PBS by gentle pipetting.
4. Place the tube on the magnetic stand to collect beads for 2–3 min and aspirate the liquid.

5. Wash the beads twice by repeating **steps 3** and **4** (wash three times in total).
6. Resuspend the beads in 250 μL of 0.5% BSA–PBS with 10 μg antibodies by gentle pipetting (*see* **Note 5**).
7. Incubate overnight on a rotating platform at 4 °C.

Day 2

Preparation of Chromatin

8. Thaw the cell pellets in 15 mL tubes on ice if the cell pellets were stored at −80 °C after Subheading 3.1, **step 26**.
9. Resuspend 10^7 cells in 5 mL cold Lysis buffer by gentle pipetting.
10. Incubate the tube at 4 °C for 10 min with mild agitation on a platform shaker.
11. Centrifuge at 600 × *g* for 5 min at 4 °C and discard the supernatant.
12. Resuspend the pellet in 6 mL Buffer 2 by gentle pipetting.
13. Place the tube at room temperature for 10 min.
14. Centrifuge at 600 × *g* for 5 min and discard the supernatant.
15. Resuspend the pellets in 1.0 mL cold Buffer 3 by gentle pipetting and keep on ice.
16. Sonicate DNA by 10 sessions of 30 pulses (1 s on and 1 s off) at 50% amplitude using a Branson Digital Sonifier System on wet ice. Let the tube sit on ice for 3 min to chill between sessions (*see* **Notes 6** and 7).
17. Add 20% N-lauroylsarcosine sodium solution at the final concentration of 0.5%.
18. Place the tube at room temperature for 10 min with mild agitation on a platform shaker.
19. Centrifuge at 1200 × *g* for 10 min at 4 °C and collect the supernatant (1 mL) as chromatin samples of 10^7 cells.
20. Reserve 2.5 μL from the chromatin samples as input controls.

Washing of Antibody-Conjugated Magnetic Beads

21. From Subheading 3.2, **step 7**: Centrifuge the magnetic beads solution at 900 × *g* for 1 min at 4 °C.
22. Place the tube on the magnetic stand to collect beads for 2–3 min and aspirate the liquid.
23. Resuspend the beads in 1 mL of 0.5% BSA–PBS by gentle pipetting (3–5 times).
24. Place the tube on the magnetic stand to collect beads for 2–3 min and aspirate the liquid.

25. Wash the beads twice by repeating **steps 23** and **24** (wash three times in total).
26. Resuspend the beads in 150 μL of 0.5% BSA–PBS by gentle pipetting (3–5 times) and keep on ice.

ChIP

27. From Subheading 3.2, **step 19**: Add 300 μL of the ChIP cocktail mix to 1 mL of the chromatin sample.
28. Add 30–150 μL of the antibody-conjugated beads solution (from **step 26**) to each sample (2–10 μg antibodies per 10^7 cells; *see* **Note 8**).
29. Incubate overnight on a rotating platform at 4 °C.

Day 3

Washing of Beads

30. From Subheading 3.2, **step 29**: Centrifuge the tube at 900 × *g* for 1 min at 4 °C.
31. Place the tube on the magnetic stand to collect beads for 2–3 min and aspirate the liquid.
32. Resuspend the beads in 1 mL of cold RIPA buffer by gentle pipetting (3–5 times) and quickly transfer to a prechilled new tube.
33. Place the tube on the magnetic stand to collect beads for 2–3 min and aspirate the liquid.
34. Wash the beads 4 times by repeating **steps 32** and **33** (5 times in total).
35. Resuspend the beads in 1 mL of cold 50 mM NaCl–TE by gentle pipetting (3–5 times) and quickly transfer to a prechilled new tube.
36. Place the tube on the magnetic stand to collect beads for 2–3 min and aspirate the liquid.

Elution

37. Resuspend the beads in 100 μL of Elution buffer by gentle pipetting.
38. Keep the tube at 65 °C for 15 min (vortex every 5 min).
39. Centrifuge at 20,000 ×x *g* for 1 min.
40. Transfer the liquid to a new tube placed on a magnetic stand.
41. Collect residual beads with a magnetic stand and transfer the liquid to a new tube.

Reverse Cross-link

42. Reverse cross-link by incubating the eluate overnight at 65 °C.

43. Add 100 μL of Elution buffer to the input control and reverse cross-link as well.

Day 4

Purification of ChIP DNA (*See* **Note 9**)

44. Add 100 μL TE to all fractions (total volume, 200 μL).
45. Add RNaseA at the final concentration of 0.2 mg/mL.
46. Incubate at 37 °C for 1 h.
47. Add Proteinase K at the final concentration of 0.2 mg/mL.
48. Incubate at 55 °C for 2 h.
49. Purify the DNA using a MinElute PCR Purification Kit according to the manufacturer's instruction and elute in 20 μL (*see* **Note 10**).
50. Measure DNA concentration using a QubitTM Fluorometer with a dsDNA HS Assay Kit.

3.3 ChIP-Seq Library Construction

- The ChIP-seq library is constructed from ChIP DNA and control DNA (from an input control or control ChIP with normal IgG) obtained in Subheading 3.2, **step 49**.
- Library construction kits or equivalent sets of reagents and equipment, which are designed for different NGS platforms, are available from various suppliers.
- Choose appropriate kits/sets and construct the libraries according to the manufacturers' instructions [19].
- Perform purification and size selection of the library by Agencourt AMPure XP according to the manufacturer's instruction (*see* **Note 11**).
- Integrity of the constructed libraries (size distribution and purity) is typically examined by an Agilent 2100 Bioanalyzer with a High Sensitivity DNA Kit.

3.4 Sequencing by NGS

Illumina NGS platforms including NextSeq, HiSeq, and NovaSeq series are commonly used. In these platforms, multiplex sequencing is conducted to allow large numbers of libraries to be pooled and sequenced simultaneously during a single run. Single-end reads are often used for ChIP-seq analysis, whereas paired-end reads could improve the mapping efficiency, especially in repeat sequence regions [20]. Requirements of sequence depth depend on the targets. In the case of ChIP-seq for transcription factors, which act as "point-source" factors providing localized sharp peaks, 20–30 million reads are required. In the case of ChIP-seq for histone markers, which act as "broad-source" factors providing large enriched domains, 40–50 million reads are required as a practical minimum [21, 22].

3.5 ChIP-Seq Data Analysis

In the Illumina NGS platform, sequencing reads are obtained as fastq files. Computational analysis of sequence data is not described in detail here. As there are several pipelines for the analysis [23], we will briefly summarize a typical set of computational analyses.

With the fastq files, the quality of sequencing reads is initially checked by programs such as FastQC [24]. After the reads are trimmed and cropped (this includes removal of adapter sequences from reads) by Trimmomatic [25] or other equivalent programs, they are mapped onto the genome using mapping tools such as Bowtie and Bowtie2 [26]. Enrichments of the mapped reads are then detected as "peaks" against background reads from input samples or control ChIP samples with normal IgG (*see* **Note 5**). This procedure is called "peak calling." A large number of programs are available for peak calling [23], including MACS [27], CisGenome [28], and so on; we have used the CisGenome platform. The mapped ChIP-seq tags are visualized on a genome browser, in which local distribution and intensity of the signals are observed. This is useful to evaluate the ChIP-seq data, especially if binding regions are already known on the genome.

Once peak regions are identified over the genome as binding profiles of factors of interest, the biological significance of those regions can be predicted by GREAT analysis, which associates the peak regions with the ontology of their putative target genes [29]. Enrichment of transcription factor motifs under the peak regions can be analyzed by de novo motif analysis; the analysis predicts transcriptional regulators working on the peak regions. Several algorithms are available for the analysis [30]; we have mainly used those in the CisGenome package [28], Homer package [31], or MEME-ChIP [32].

In addition to those, further analysis can provide mechanistic insight into the action of transcription factors of interest. For example, binding regions of different factors are often compared by the intersection of peaks using bedtools [33]. Such comparison reveals not only binding overlap between the factors, but also mutually exclusive regions in their binding profiles. By taking advantage of an algorithm used in RNA-seq, statistical analysis can be performed to identify differentially bound sites [34]. In the RNA-seq platform, correlation analysis of multiple binding profiles can be tested for similarity among the profiles. Lastly, integrative analysis of DNA binding profiles in ChIP-seq and gene expression profiles in RNA-seq or microarray is a powerful approach to identify target genes for transcription factors of interest. By narrowing down the candidates through such analysis, we have only to test a limited number of loci for their biological significance by wet experiments including reporter analysis and functional assays.

4 Notes

1. Calcified and uncalcified cartilage regions can be distinguished, since the former looks whiter than the latter. Cut on the boundary under the microscope to collect uncalcified cartilage.
2. When ChIP is performed on cells cultured in vitro, collect the cells in tubes and start from this step.
3. We typically use 10^7 cells per ChIP. Thus a convenient number of cells per tube would be 10^7. However, the number of cells per ChIP must be optimized depending on the types of cells, performance of antibodies, and so on.
4. Length of cross-linking time needs to be optimized. It can be extended up to 30 min.
5. Magnetic beads incubated with normal IgG are also prepared for control ChIP.
6. The conditions of sonication, including frequency, time, and amplitude, need to be optimized depending on the types of cells. The conditions shown were optimized for a Branson Digital Sonifier 250. Another sonication platform could also be used under optimized conditions. The recommended size of fragmented DNA is 150–500 bp.
7. If more than 10^7 cells are processed in 1 mL of Buffer 3, add additional Buffer 3 to obtain a solution containing the chromatin of 10^7 cells per 1 mL. A total 1 mL of chromatin samples from 10^7 cells is used per ChIP.
8. The amount of antibodies used per ChIP needs to be optimized.
9. For convenience, samples after reverse cross-link can be stored at −30 °C until use.
10. Among 20 μL eluate from ChIP, 1–2 μL is subjected to measurement of the concentration of ChIP DNA using a Qubit™ Fluorometer with a dsDNA HS Assay Kit, and up to 5 μL is used to test the enrichment of ChIP DNA by quantitative polymerase chain reaction (PCR) analysis.
11. The expected library size is from 200 bp to 600 bp. Excess primers at 70–75 bp and the adaptor dimer at 125–130 bp should be removed.

Acknowledgments

We thank Dr. Andrew P. McMahon and members of the McMahon lab for their helpful input on the optimization of our ChIP-seq protocols. A series of ChIP-seq studies in skeletal cells, which allowed us to develop the protocol in this chapter, were supported by a grant from the NIH (DK056246) and by JSPS KAKENHI (23689079, 25670851, and 26713054).

References

1. Barski A, Cuddapah S, Cui K, Roh TY, Schones DE, Wang Z, Wei G, Chepelev I, Zhao K (2007) High-resolution profiling of histone methylations in the human genome. Cell 129 (4):823–837. https://doi.org/10.1016/j.cell.2007.05.009
2. Visel A, Bristow J, Pennacchio LA (2007) Enhancer identification through comparative genomics. Semin Cell Dev Biol 18 (1):140–152. https://doi.org/10.1016/j.semcdb.2006.12.014
3. Visel A, Blow MJ, Li Z, Zhang T, Akiyama JA, Holt A, Plajzer-Frick I, Shoukry M, Wright C, Chen F, Afzal V, Ren B, Rubin EM, Pennacchio LA (2009) ChIP-seq accurately predicts tissue-specific activity of enhancers. Nature 457 (7231):854–858. https://doi.org/10.1038/nature07730
4. Kharchenko PV, Alekseyenko AA, Schwartz YB, Minoda A, Riddle NC, Ernst J, Sabo PJ, Larschan E, Gorchakov AA, Gu T, Linder-Basso D, Plachetka A, Shanower G, Tolstorukov MY, Luquette LJ, Xi R, Jung YL, Park RW, Bishop EP, Canfield TK, Sandstrom R, Thurman RE, MacAlpine DM, Stamatoyannopoulos JA, Kellis M, Elgin SC, Kuroda MI, Pirrotta V, Karpen GH, Park PJ (2011) Comprehensive analysis of the chromatin landscape in Drosophila melanogaster. Nature 471 (7339):480–485. https://doi.org/10.1038/nature09725
5. Rada-Iglesias A, Bajpai R, Swigut T, Brugmann SA, Flynn RA, Wysocka J (2011) A unique chromatin signature uncovers early developmental enhancers in humans. Nature 470 (7333):279–283. https://doi.org/10.1038/nature09692
6. Oh CD, Lu Y, Liang S, Mori-Akiyama Y, Chen D, de Crombrugghe B, Yasuda H (2014) SOX9 regulates multiple genes in chondrocytes, including genes encoding ECM proteins, ECM modification enzymes, receptors, and transporters. PLoS One 9(9): e107577. https://doi.org/10.1371/journal.pone.0107577
7. Hakelien AM, Bryne JC, Harstad KG, Lorenz S, Paulsen J, Sun J, Mikkelsen TS, Myklebost O, Meza-Zepeda LA (2014) The regulatory landscape of osteogenic differentiation. Stem Cells 32(10):2780–2793. https://doi.org/10.1002/stem.1759
8. Wu H, Whitfield TW, Gordon JA, Dobson JR, Tai PW, van Wijnen AJ, Stein JL, Stein GS, Lian JB (2014) Genomic occupancy of Runx2 with global expression profiling identifies a novel dimension to control of osteoblastogenesis. Genome Biol 15(3):R52. https://doi.org/10.1186/gb-2014-15-3-r52
9. Meyer MB, Benkusky NA, Pike JW (2014) The RUNX2 cistrome in osteoblasts: characterization, down-regulation following differentiation, and relationship to gene expression. J Biol Chem 289(23):16016–16031. https://doi.org/10.1074/jbc.M114.552216
10. Liu CF, Lefebvre V (2015) The transcription factors SOX9 and SOX5/SOX6 cooperate genome-wide through super-enhancers to drive chondrogenesis. Nucleic Acids Res 43 (17):8183–8203. https://doi.org/10.1093/nar/gkv688
11. Ohba S, He X, Hojo H, McMahon AP (2015) Distinct transcriptional programs underlie Sox9 regulation of the mammalian chondrocyte. Cell Rep 12(2):229–243. https://doi.org/10.1016/j.celrep.2015.06.013
12. Hojo H, Ohba S, He X, Lai LP, McMahon AP (2016) Sp7/Osterix is restricted to bone-forming vertebrates where it acts as a dlx co-factor in osteoblast specification. Dev Cell 37(3):238–253. https://doi.org/10.1016/j.devcel.2016.04.002
13. He X, Ohba S, Hojo H, McMahon AP (2016) AP-1 family members act with Sox9 to promote chondrocyte hypertrophy. Development 143(16):3012–3023. https://doi.org/10.1242/dev.134502
14. Thompson B, Varticovski L, Baek S, Hager GL (2016) Genome-wide chromatin landscape transitions identify novel pathways in early commitment to osteoblast differentiation. PLoS One 11(2):e0148619. https://doi.org/10.1371/journal.pone.0148619
15. Yamashita S, Kataoka K, Yamamoto H, Kato T, Hara S, Yamaguchi K, Renard-Guillet C, Katou Y, Shirahige K, Ochi H, Ogino H, Uchida T, Inui M, Takada S, Shigenobu S, Asahara H (2019) Comparative analysis demonstrates cell type-specific conservation of SOX9 targets between mouse and chicken. Sci Rep 9(1):12560. https://doi.org/10.1038/s41598-019-48979-4

16. Odom DT, Zizlsperger N, Gordon DB, Bell GW, Rinaldi NJ, Murray HL, Volkert TL, Schreiber J, Rolfe PA, Gifford DK, Fraenkel E, Bell GI, Young RA (2004) Control of pancreas and liver gene expression by HNF transcription factors. Science 303 (5662):1378–1381. https://doi.org/10.1126/science.1089769
17. Vokes SA, Ji H, Wong WH, McMahon AP (2008) A genome-scale analysis of the cis-regulatory circuitry underlying sonic hedgehog-mediated patterning of the mammalian limb. Genes Dev 22(19):2651–2663. https://doi.org/10.1101/gad.1693008
18. Vokes SA, Ji HK, McCuine S, Tenzen T, Giles S, Zhong S, Longabaugh WJR, Davidson EH, Wong WH, McMahon AP (2007) Genomic characterization of Gli-activator targets in sonic hedgehog-mediated neural patterning. Development 134(10):1977–1989. https://doi.org/10.1242/Dev.001966
19. Sundaram AY, Hughes T, Biondi S, Bolduc N, Bowman SK, Camilli A, Chew YC, Couture C, Farmer A, Jerome JP, Lazinski DW, McUsic A, Peng X, Shazand K, Xu F, Lyle R, Gilfillan GD (2016) A comparative study of ChIP-seq sequencing library preparation methods. BMC Genomics 17(1):816. https://doi.org/10.1186/s12864-016-3135-y
20. Chen Y, Negre N, Li Q, Mieczkowska JO, Slattery M, Liu T, Zhang Y, Kim TK, He HH, Zieba J, Ruan Y, Bickel PJ, Myers RM, Wold BJ, White KP, Lieb JD, Liu XS (2012) Systematic evaluation of factors influencing ChIP-seq fidelity. Nat Methods 9 (6):609–614. https://doi.org/10.1038/nmeth.1985
21. Bailey T, Krajewski P, Ladunga I, Lefebvre C, Li Q, Liu T, Madrigal P, Taslim C, Zhang J (2013) Practical guidelines for the comprehensive analysis of ChIP-seq data. PLoS Comput Biol 9(11):e1003326. https://doi.org/10.1371/journal.pcbi.1003326
22. Jung YL, Luquette LJ, Ho JW, Ferrari F, Tolstorukov M, Minoda A, Issner R, Epstein CB, Karpen GH, Kuroda MI, Park PJ (2014) Impact of sequencing depth in ChIP-seq experiments. Nucleic Acids Res 42(9):e74. https://doi.org/10.1093/nar/gku178
23. Nakato R, Shirahige K (2017) Recent advances in ChIP-seq analysis: from quality management to whole-genome annotation. Brief Bioinform 18(2):279–290. https://doi.org/10.1093/bib/bbw023
24. Yang X, Liu D, Liu F, Wu J, Zou J, Xiao X, Zhao F, Zhu B (2013) HTQC: a fast quality control toolkit for Illumina sequencing data. BMC Bioinformatics 14:33. https://doi.org/10.1186/1471-2105-14-33
25. Bolger AM, Lohse M, Usadel B (2014) Trimmomatic: a flexible trimmer for Illumina sequence data. Bioinformatics 30 (15):2114–2120. https://doi.org/10.1093/bioinformatics/btu170
26. Langmead B, Trapnell C, Pop M, Salzberg SL (2009) Ultrafast and memory-efficient alignment of short DNA sequences to the human genome. Genome Biol 10(3):R25. https://doi.org/10.1186/gb-2009-10-3-r25
27. Zhang Y, Liu T, Meyer CA, Eeckhoute J, Johnson DS, Bernstein BE, Nusbaum C, Myers RM, Brown M, Li W, Liu XS (2008) Model-based analysis of ChIP-Seq (MACS). Genome Biol 9(9):R137. https://doi.org/10.1186/gb-2008-9-9-r137
28. Ji H, Jiang H, Ma W, Johnson DS, Myers RM, Wong WH (2008) An integrated software system for analyzing ChIP-chip and ChIP-seq data. Nat Biotechnol 26(11):1293–1300. https://doi.org/10.1038/nbt.1505
29. McLean CY, Bristor D, Hiller M, Clarke SL, Schaar BT, Lowe CB, Wenger AM, Bejerano G (2010) GREAT improves functional interpretation of cis-regulatory regions. Nat Biotechnol 28(5):495–501. https://doi.org/10.1038/nbt.1630
30. Boeva V (2016) Analysis of genomic sequence motifs for deciphering transcription factor binding and transcriptional regulation in eukaryotic cells. Front Genet 7:24. https://doi.org/10.3389/fgene.2016.00024
31. Heinz S, Benner C, Spann N, Bertolino E, Lin YC, Laslo P, Cheng JX, Murre C, Singh H, Glass CK (2010) Simple combinations of lineage-determining transcription factors prime cis-regulatory elements required for macrophage and B cell identities. Mol Cell 38 (4):576–589. https://doi.org/10.1016/j.molcel.2010.05.004
32. Machanick P, Bailey TL (2011) MEME-ChIP: motif analysis of large DNA datasets. Bioinformatics 27(12):1696–1697. https://doi.org/10.1093/bioinformatics/btr189
33. Quinlan AR, Hall IM (2010) BEDTools: a flexible suite of utilities for comparing genomic features. Bioinformatics 26(6):841–842. https://doi.org/10.1093/bioinformatics/btq033
34. Stark R, Brown G (2019) DiffBind: differential binding analysis of ChIPSeq peak data

Chapter 13

MicroRNAs and Regulation of Autophagy in Chondrocytes

Yolande F. M. Ramos and Ali Mobasheri

Abstract

Chondrocytes are the main cells responsible for the maintenance of cartilage homeostasis and integrity. During development, extracellular matrix (ECM) macromolecules are produced and deposited by chondrocyte precursors. Autophagy, a highly dynamic process aimed at degradation of dysfunctional or pathogenic proteins, organelles, and intracellular microbes that can damage tissues, is one of the key processes required for sustained cartilage homeostasis. In different cell types it has been shown that, among others, autophagy is regulated by epigenetic mechanisms such as small noncoding RNAs (miRNAs, ~22 base pairs). Increasing evidence suggests that miRNAs are also involved in the regulation of autophagy in chondrocytes. Based on our previous research of gene and miRNA expression in articular cartilage, in this chapter we provide a summary of the tools models to direct in vitro and in vivo studies aimed at gaining a better understanding of the regulatory roles of miRNAs in chondrocyte autophagy.

Key words Chondrocyte, microRNA, Autophagy, Cartilage, Extracellular matrix, Osteoarthritis, Autophagy interactome

1 Introduction

During embryonic development, cartilage extracellular matrix (ECM) is produced deposited by chondrocyte precursors. Into adulthood, articular cartilage covers and protects the ends of (long) bones to ensure low-friction joint articulation and smooth body movement by virtue of its flexibility, elasticity, and load-bearing properties. These unique characteristics of articular cartilage result from the specialized ECM, predominantly consisting of type II collagen fibers (7.6%) and polyanionic proteoglycans (11%) with the capacity to draw large quantities of cations (i.e., Na^+, K^+, and Ca^{2+}) into the matrix, allowing it to swell and retain a high water content (60%) [1].

Throughout life, cartilage homeostasis and integrity is maintained by the terminally differentiated and highly specialized chondrocytes, the sole cell population in cartilage ECM. Following (micro)traumas in cartilage ECM, emerging from wear and tear, chondrocytes need to rapidly turn on and off specific genes to

Tariq M. Haqqi and Véronique Lefebvre (eds.), *Chondrocytes: Methods and Protocols*, Methods in Molecular Biology, vol. 2245, https://doi.org/10.1007/978-1-0716-1119-7_13, © Springer Science+Business Media, LLC, part of Springer Nature 2021

restore the damage, remove breakdown products, and deposit new matrix. To control and certify such changes in gene expression and conserve tissue homeostasis, the maturational arrested chondrocytes heavily depend on epigenetic mechanisms [2, 3]. Epigenetic processes affect gene expression beyond mere genetic makeup. Aging leads to loss of epigenetic control mechanisms, resulting in aberrant gene expression with concomitant loss of homeostasis. Eventually, this results in osteoarthritis (OA), the predominant form of joint disease, characterized by loss of cartilage integrity causing pain and disability in patients [4]. Despite increasing prevalence of OA, the precise etiology underlying its onset and development remains largely unknown resulting in poor treatment options. Improved insight into fundamental biological processes underlying loss of chondrocyte homeostasis will eventually advance and inform the development of new therapeutic strategies, leading to better and more effective medications for this serious disease.

1.1 MicroRNA: Biogenesis and Role in Chondrogenesis

Different types of epigenetic mechanisms controlling gene expression and translation are recognized including DNA methylation, histone modification, and noncoding RNAs. Noncoding RNA molecules are classified into long noncoding RNAs (lncRNAs) with at least 200 nucleotides, and small noncoding RNAs of around 22 base pairs in size. Among these small noncoding RNAs, microRNAs (miRNAs) have been most extensively studied [3].

The majority of miRNAs localize to introns, but some miRNAs that localize to exons have also been identified [5]. The process from transcription toward mature noncoding miRNAs is complex (Fig. 1a). In the nucleus of a cell, miRNAs are transcribed as hairpin-containing primary transcripts (pri-miRNAs). Subsequently, pri-miRNAs are cleaved into pre-miRNAs of around 100 nucleotides by a protein named Drosha, and transported to the cytoplasm by Exportin-5 where protein called Dicer further cleaves the pre-miRNAs into double stranded miRNAs. To become functionally active, the double stranded molecule is unwound by helicases into single-stranded miRNAs, and either the 3-prime (3p) or the 5-prime (5p) miRNA is stabilized by virtue of its binding with Argonaute (AGO) protein to form the microRNA-induced silencing complex (mRISC) [5].

Since the first discovery of miRNAs more and more miRNAs have been identified. As of October 2019, according to GENCODE data (v.32) 1818 human miRNAs are known while miRBase catalogue (v22) even contained 1917 annotated pri-miRNAs and 2654 mature sequences. Expression levels of miRNAs are highly variable. In this respect, by applying deep sequencing of miRNAs Haseeb and colleagues showed that in chondrocytes limited number of miRNAs are expressed at high levels: out of 313 miRNAs detected in primary chondrocytes, twenty miRNAs represented about 85% of all reads [6].

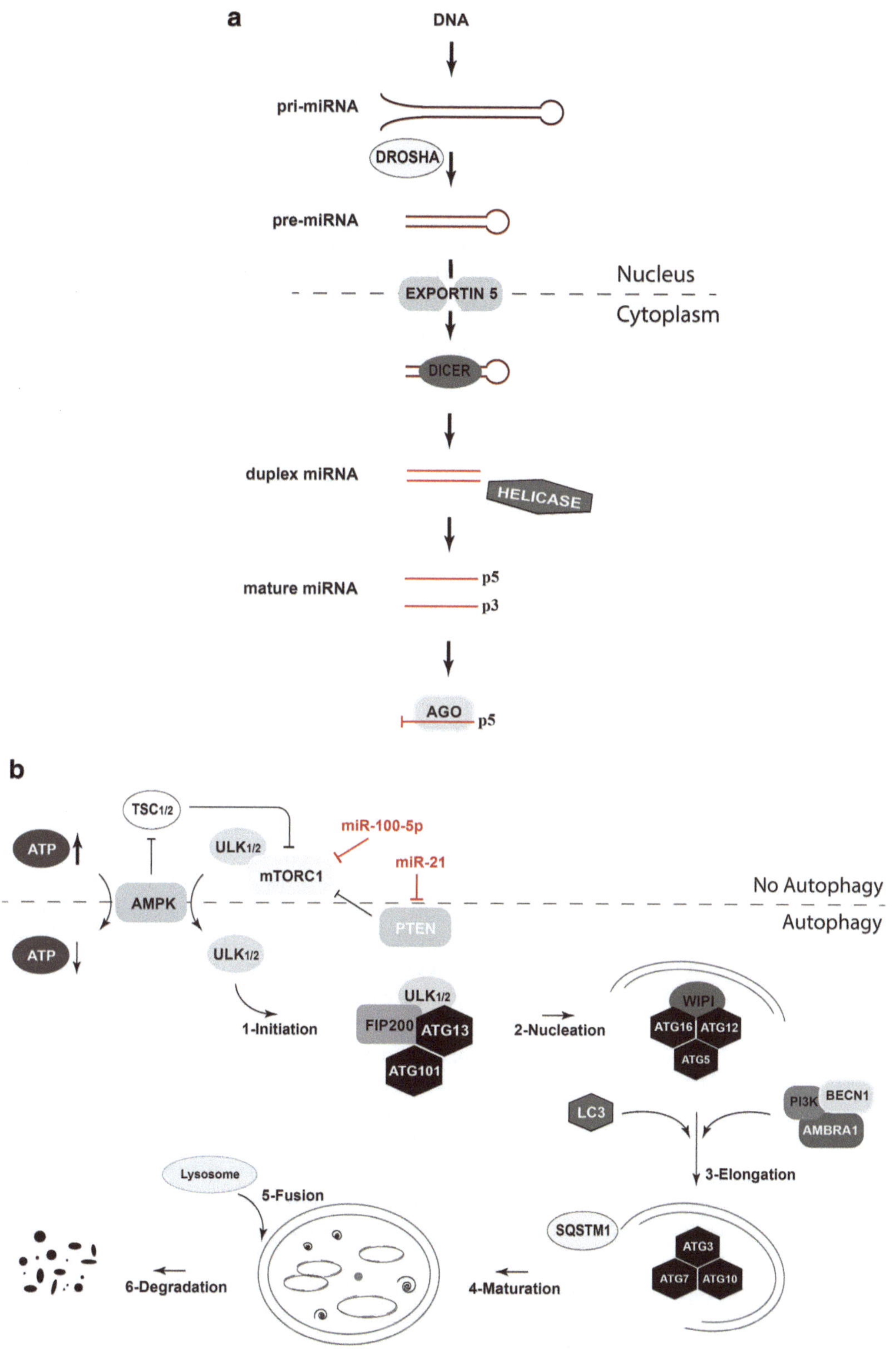

Fig. 1 Schematic representation of (**a**) miRNA biogenesis and (**b**) sequential steps of autophagy; with selected essential proteins involved in each successive step

MicroRNAs are involved in the regulation of a wide range of biological processes, from proliferation to differentiation and homeostasis. An essential role for miRNAs in skeletogenesis and chondrogenesis was demonstrated by knocking out Dicer [7] or Drosha [8], resulting in disrupted expression of the majority of the miRNAs. Mice without Dicer show progressive reduction in the proliferating pool of chondrocytes and concurrent severe skeletal growth defects leading to premature death. Likewise, postnatal disruption of Drosha expression in mice showed a proliferation defect and death in articular chondrocytes. Specific miRNAs playing a role in cartilage health and disease have also been identified and an informative overview is presented, among others, by Swingler et al. [5] discussing the function of multiple miRNAs.

1.2 Autophagy in Chondrocytes

Among the different biological pathways regulated by miRNAs, recently several laboratories have shown that autophagy is also regulated by miRNAs. Autophagy is a highly dynamic process used by eukaryotes to specifically degrade dysfunctional or pathogenic proteins, organelles, and intracellular microbes that can damage tissues [9, 10]. Therefore, autophagy is an essential cellular mechanism to protect from cell death and warrant tissue integrity, including cartilage homeostasis [11].

The process of autophagy consists of multiple steps with many different proteins playing key roles (Fig. 1b). Association of mammalian target of rapamycin (mTOR) with uncoordinated 51-like kinase (ULK) family members, prevents autophagy. Following release from mTOR as a result from adverse events such as deprivation of nutrients and energy, ULK1 and ULK2 complex together with FIP200 and several autophagy-related genes (ATGs) to initiate formation of a phagophore [10], Upon association with WIPI (WD repeat domain phosphoinositide-interacting) proteins and subsequent conjugation with ubiquitin-like (Ub-like) proteins ATG12, ATG5, and ATL16L1, followed by sequential binding and release of other ATG-family members, elongation and maturation of the autophagophore is established. Essential in this respect is also presence of the Ub-like LC3 subfamily. In fact, many studies addressing incidence of autophagy show changes in expression of MAP 1LC3-I (LC3-I) and LC3-II by Western blotting and/or immunofluorescence to demonstrate incidence of autophagy, being LC3-II the phosphatidylethanolamine-conjugated form of MAP 1LC3 frequently used as marker for autophagy. [12] Finally, fusion of the phagophore with a lysosome leads to degradation of its content. The different components of the phagophore are released and recycled.

Upon aging and cellular senescence, levels of autophagy are reduced promoting pathogenesis, as seen in chondrocytes from OA joints [13]. In line with this, both in vitro and in vivo experimental models have demonstrated that restoration of autophagy could support homeostasis and life span [14, 15], and protect from the development of OA [16].

Although many lines of evidence have suggested a role for miRNAs in the regulation of autophagy, to our knowledge it was not until 2014 when Song and colleagues [17] actually presented evidence for a relation between expression of miRNAs and autophagy in chondrocytes. In their study, Song et al. showed that introduction of miR-21 inhibited the expression levels of autophagic complexes in chondrocytes via modulation of ATG levels which in vivo resulted in significantly increased cartilage matrix destruction. However, direct regulation of autophagy via miRNAs in chondrocytes was not demonstrated until recently. To this end, binding between miRNAs and specific autophagy-related mRNAs was shown in chondrocytes by applying luciferase assays. Taking this approach, for example, miR-21 was shown to directly bind to the 3′ untranslated region (3′UTR) of phosphatase and tensin homolog (*PTEN*) and inhibit its expression and subsequently the incidence of autophagy in primary nucleus pulposus cells [18]. Likewise, miR-100-5p and miR-27a were suggested to decrease autophagy by direct binding and inhibition of *mTOR* 3′UTR [19] and *PI3K* [20], respectively, both determined with luciferase assays, though not in chondrocytes but in SW1353 chondrosarcoma cells. In addition to miR-21, miR-27a, and miR-100-5p, the roles of numerous other miRNAs in chondrocyte autophagy have been studied, both in vitro and in vivo (reviewed in [5, 21]), although frequently without showing direct association with the mRNA.

1.3 Interactions of miRNA with Genes Regulating Autophagy: The Autophagy Interactome

It is known that a single miRNA can target multiple mRNAs. The specificity of the interactions is ensured by 6–8 nucleotides, the so-called seed sequences of miRNAs, mostly matching a sequence in the 3′UTR of an mRNA molecule. Upon association of miRNA with mRNA molecules they post-transcriptionally diminish protein expression via degradation of the mRNA and inhibition of translation [2, 3]. Frequently, miRNAs share promoters and concurrent regulation of expression with their host gene. Furthermore, regulation of gene expression is highly tissue specific, and also it is becoming more and more clear that miRNA expression with expression of its targets is tissue-specific as exemplified recently by Deng et al. [22] and by Zhou et al. [23] Taken together, this renders studying the interaction and regulation of miRNAs and its target-gene expression particularly challenging, and thus a systems biology approach in the appropriate tissue is recommended [24]. Subsequently, to formerly prove a regulatory role for specific miRNAs in chondrocyte-autophagy we emphasize the importance of demonstrating binding of the miRNA to mRNA, preferentially in appropriate cell model given that epigenetic regulation is highly cell type specific. When investigating expression of miRNAs in primary cells, it should be taken into consideration that miRNAs play a role in cell proliferation. Its expression may be considerably different when comparing expression in tissues (here: autologous articular cartilage) with that in proliferating cells in culture (here: primary chondrocytes on plastic in 2D while expanding).

To direct future studies with respect to miRNA-regulated autophagy in chondrocytes, a literature-based selection was made of 47 genes that are generally known to be involved in autophagy in different tissues, including transcription factors nuclear factor NF-Kappa-B (*NFκB*) 1 and 2 (Table 1). Expression levels of these genes in articular cartilage as determined by RNA-sequencing were correlated with miRNA expression levels in the same samples ($N = 19$) measured with miRNA-sequencing as available from the RAAK study in our laboratory [24, 25]. Analysis resulted in 220 inverse correlations of at least 0.60 between miRNAs and mRNAs. Subsequent selection for opposite direction of effects to identify robust and likely functional correlations identified 15 unique miRNAs that correlate with 4 unique genes (Fig. 2 and Table 2). Most significant was the correlation between miR-4510 and *MAP 1LC3B* ($r = -0.84$; $P = 8.0 \times 10^{-6}$). During OA, miR-4510 is 1.3-fold downregulated in articular cartilage ($P = 3.2 \times 10^{-2}$) while expression of *MAP 1LC3B* is 1.3-fold increased (i.e., lesioned compared to preserved OA cartilage; $P = 3.8\text{x}10^{-3}$). Although expression of miR-155 was previously found to modulate autophagy via *MAP 1LC3B* [26], in our articular cartilage samples we did not find significant inverse correlation between miR-155 and *MAP 1LC3B* ($r = -0.29$; $P = 2.3 \times 10^{-1}$). More importantly, expression of both *MAP 1LC3B* and miR-155 (1.8-fold; $P = 9.5 \times 10^{-4}$) is significantly increased when comparing lesioned with preserved OA cartilage while effects in the opposite direction are expected in case an miRNA regulates expression of a particular gene. A well-known and highly expressed miRNA in chondrocytes is miR-140-3p that inversely correlated to *WIPI1* ($r = -0.65$; $P = 2.7 \times 10^{-3}$). Expression of both, miR-140-3p as well as *WIPI1*, change significantly during OA (respectively, 1.3-fold down with $P = 3.1 \times 10^{-3}$ and 1.3-fold up with $P = 6.3 \times 10^{-6}$). Regulation of *WIPI1* by miR-140-3p, however, has not been shown to date. As such, identified correlations are attractive candidates for further studies to gain knowledge of previously unknown miRNA targets to restore and regulate autophagic mechanisms and pathways in chondrocytes in OA.

Table 1
Literature-based selection of genes involved in autophagy that were correlated with miRNA expression levels in articular cartilage

AMBRA1	PIK3CA
ATG10	PIK3CB
ATG101	PIK3CD
ATG12	PIK3CDP1
ATG13	PIK3CG
ATG16L1	PIK3IP1
ATG16L2	PRKAB1
ATG3	PRKAB2
ATG5	PTEN
ATG7	RB1CC1
BECN1	SMARCA2
BECN2	SMARCA4
DDIT4	SQSTM1
MAP1LC3A	TSC1
MAP1LC3B	TSC2
MAP1LC3B2	ULK1
MAP1LC3C	ULK2
MAP1LC3P	ULK3
MTOR	ULK4
PIK3AP1	WIPI1
PIK3C2A	WIPI2
PIK3C2B	NFKB1
PIK3C2G	NFKB2
PIK3C3	

2 Materials

2.1 Culture Media

1. Phosphate buffered saline (PBS) 0.9%, pH 7.4.
2. DMEM Medium, high glucose, pyruvate added.
3. Fetal bovine serum (FBS).
4. Penicillin–Streptomycin.
5. Basic fibroblast growth factor (bFGF-2).

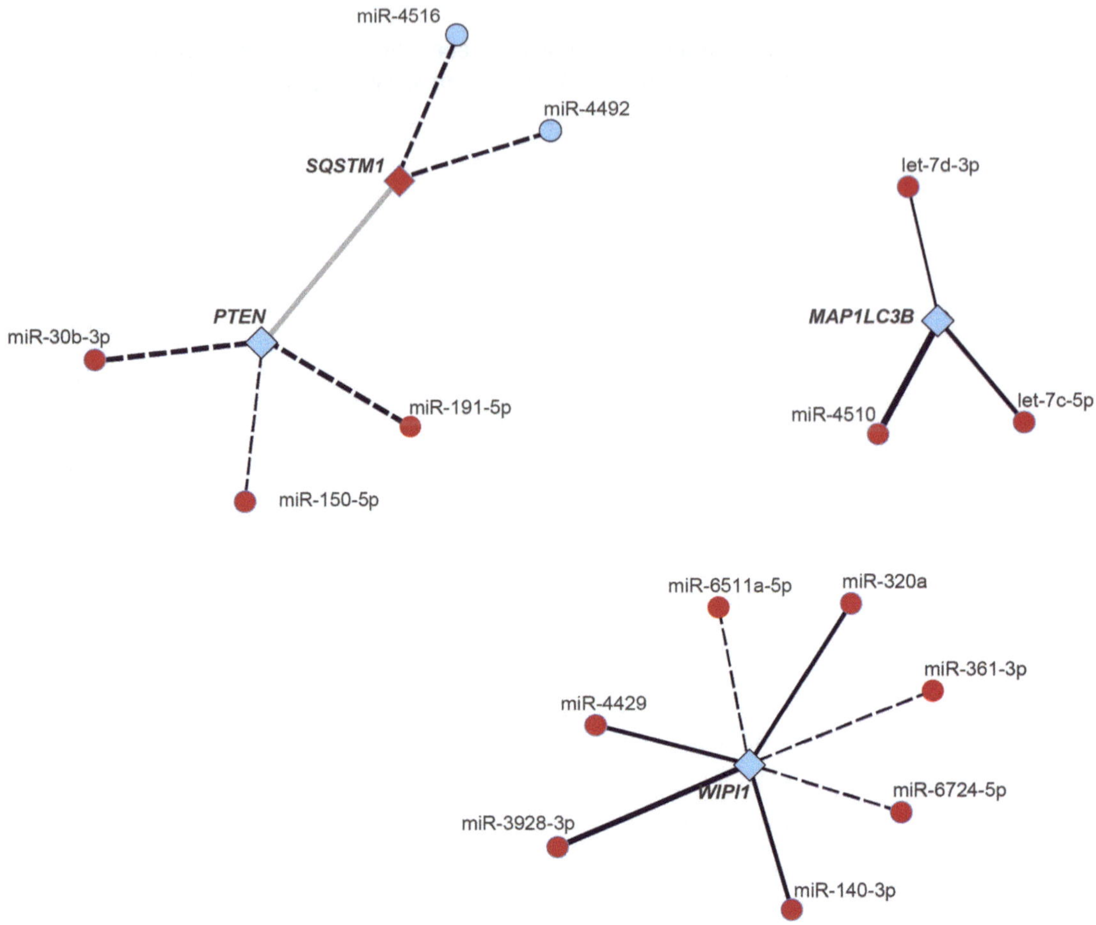

Fig. 2 Cartilage Autophagy Interactome with inverse correlations between miRNA-mRNA in articular cartilage of patients undergoing total joint replacement surgery (legend: circle: miRNA; diamond: mRNA; red: upregulation and blue: downregulation in lesioned OA cartilage as compared to preserved OA cartilage; dashed line: miRNA expression changes not significant)

6. Collagenase type I.
7. Trypsin–EDTA.
8. Opti-MEM I reduced serum medium.

2.2 Chemicals and Kits

1. Lipofectamine Transfection Reagent.
2. miRNA mimics or inhibitors from mirVana.
3. TRIzol Reagent.
4. Chloroform.
5. miRNA isolation Kit.
6. Ethanol (100%; 70%).
7. 3 M sodium acetate (pH 5.2).

Table 2
Correlations cartilage autophagy interactome

		miRNA-mRNA		miRNA (P vs L)		mRNA (P vs L)	
miRNA	**mRNA**	**corr.**	**Pval**	**Pval**	**FC**	**FC**	**Pval**
miR-4510	*MAP1LC3B*	−0.84	8.0E-06	3.2E-02	0.76	1.25	3.8E-03
miR-3928-3p	*WIPI1*	−0.73	4.2E-04	3.8E-02	0.69	1.31	6.3E-06
miR-30b-3p	*PTEN*	−0.70	9.4E-04	4.3E-01	0.85	1.22	3.1E-03
miR-4429	*WIPI1*	−0.67	1.7E-03	7.8E-03	0.58	1.31	6.3E-06
miR-191-5p	*PTEN*	−0.68	1.3E-03	4.1E-01	0.96	1.22	3.1E-03
let-7c-5p	*MAP1LC3B*	−0.67	1.9E-03	4.8E-01	0.9	1.25	3.8E-03
miR-4492	*SQSTM1*	−0.67	1.9E-03	1.7E-01	1.43	0.81	1.2E-04
miR-4516	*SQSTM1*	−0.66	2.3E-03	1.8E-01	1.28	0.81	1.2E-04
miR-140-3p	*WIPI1*	−0.65	2.7E-03	3.1E-03	0.81	1.31	6.3E-06
miR-361-3p	*WIPI1*	−0.64	2.9E-03	2.3E-01	0.92	1.31	6.3E-06
miR-320a	*WIPI1*	−0.64	3.1E-03	9.6E-03	0.66	1.31	6.3E-06
miR-6511a-5p	*WIPI1*	−0.63	4.1E-03	6.3E-01	0.89	1.31	6.3E-06
miR-6724-5p	*WIPI1*	−0.62	4.3E-03	5.7E-01	0.82	1.31	6.3E-06
let-7d-3p	*MAP1LC3B*	−0.62	4.4E-03	7.0E-01	0.94	1.25	3.8E-03
miR-150-5p	*PTEN*	−0.62	4.8E-03	8.1E-01	0.96	1.22	3.1E-03
PTEN	*SQSTM1*	−0.77	1.1E-04	NA	NA	0.81	1.2E-04

Legend: corr.: miRNA – mRNA correlation coefficient; FC: fold change expression levels in lesioned versus preserved OA cartilage

8. cDNA Synthesis Kit for miRNA.
9. qPCR Master mix.
10. TaqMan assays.

3 Methods

3.1 Chondrocyte Culture

1. Prepare medium (DMEM, supplemented with 10% FBS, 20 U/ml pent-strep, 0.5 ng/ml bFGF-2) and prewarm at 37 °C.
2. Wash joint tissues with PBS (room temperature) and explore to select macroscopically unaffected area for collection of intact cartilage areas (preferably of weight-bearing area) (*see* **Note 1**).
3. Transfer cartilage to a Ø 10 cm culture dish and cut into very small pieces (approx. 1 mm^2) using a sterile scalpel.

4. Add medium with collagenase (2 mg/ml) onto the pieces of cartilage and distribute evenly in the dish.
5. Leave the dish overnight (~16 h) in a humidified 37 °C incubator with 5% CO_2/95% atmosphere (37 °C, 5% CO_2).
6. Next day, place a cell strainer (70 μm pore size) on a 50 ml tube.
7. Carefully collect the medium with collagenase-treated cartilage using a 10 ml pipette, and pipette the cell suspension through the cell strainer (do not overflow the strainer) to remove undigested cartilage fragments and debris.
8. Collect chondrocytes by centrifugation of the 50 ml tube (5 min at 295 × *g*).
9. Prepare a new Ø 10 cm dishes and seed primary chondrocytes after resuspension in 10 ml medium. Make sure cells are well-distributed on the dish. Expand cells until semiconfluency (~90%; Fig. 3) (*see* **Note 2**).

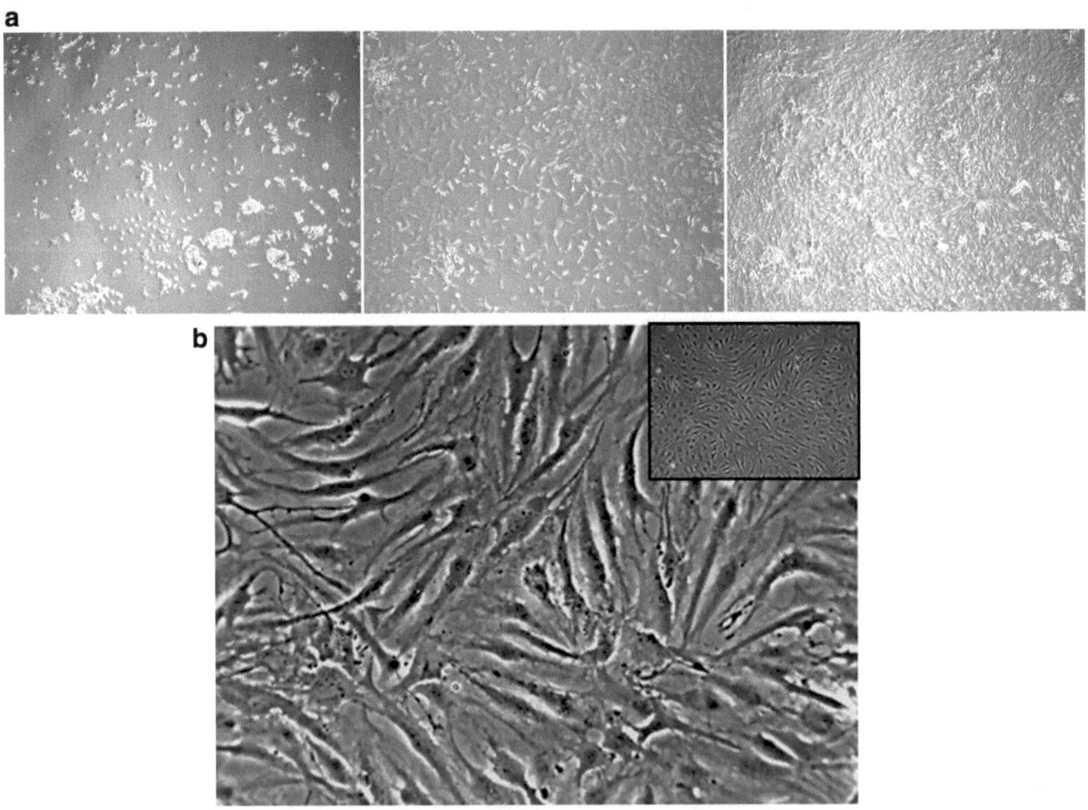

Fig. 3 Primary articular chondrocytes. (**a**) Proliferating chondrocytes at first passage following isolation from cartilage, 1 day following passaging (left panel; sparse density), ~3 days following passaging (middle panel; 80–90% confluency), and 4–5 days following passaging (right panel; confluent). (**b**) Higher magnification of confluent chondrocytes

3.2 Transfection of miRNA

1. Remove medium and wash cells once with EDTA.
2. Add 1 ml trypsin–EDTA, and incubate cells at 37 °C for ~5 min.
3. Check cells under microscope and upon confirmation of detachment, resuspend cells in medium.
4. Count cells (*see* **Note 3**), and seed chondrocytes in 24-well plate using 4 × 10^4 cells per well in 0.5 ml medium, and place back in incubator (37 °C, 5% CO2). Prepare one well per condition to be analyzed including controls (miRNA mimic, miRNA inhibitor, aspecific miRNA mimic and inhibitor, mock).
5. Next day (at least 16 h later), prepare transfection mixtures according to manufacturer's instructions.

   In our hands, good transfection efficiencies for small RNAs in chondrocytes were obtained using Lipofectamine RNAi-MAX Transfection Reagent (ThermoFisher):

   (a) Add 1 μl Lipofectamine RNAiMAX to 50 μl Opti-MEM.

   (b) Add 2 μl (3 μM) miRNA mimic/inhibitor to 50 μl Opti-MEM.

   (c) Combine mixture a and b by gentle pipetting and incubate at room temperature for 15 min.

   (d) Add all (100 μl total) to a 24-well and mix gently (the final volume in the well is now 600 μl).
6. Incubate cells for 24–48 h in incubator (37 °C, 5% CO_2).
7. To collect RNA, wash cells with 500 μl ice-cold PBS and add 300 μl TRIzol per well. Resuspend lysate thoroughly in well, collect, and store at −80 °C in Eppi tube until RNA extraction.

3.3 Quantifying miRNA Expression

1. Add 60 μl (0.2× volume) chloroform to cell lysates in TRIzol, mix well by inverting (do **not vortex**!), and incubate at room temperature for 5 min to dissociate nucleoprotein complexes.
2. Centrifugate at 18,407 × *g* (10 min, room temperature), and transfer the (transparent) upper phase to an RNase-free tube. Do not touch the interphase: contamination with proteins and/or TRIzol will decrease efficiency of downstream procedures.
3. Perform miRNA isolation using miRNA isolation Kit according to manufacturer's protocol (*see* **Note 4**), and elute miRNA in 30–50 μl RNase-free water (*see* **Note 5**).
4. Determine concentration with NanoDrop or Qubit and check RNA Integrity Number (RIN) using 2100 BioAnalyzer (Agilent) with RNA 6000 Nano Kit (*see* **Note 6**).

Quantification of miRNA expression can be done genome wide (miRNA sequencing or miRNAseq) or target specific (RT-qPCR).

When applying miRNAseq library preparations should be performed. Advantages of miRNAseq are that it allows for reliable quantitative comparison across the different miRNAs and also of low expressed miRNAs (*see* **Note 7**).

When applying RT-qPCR, cDNA should be synthesized. Advantages are the lower costs and speed of methods as compared to miRNAseq. However, for each miRNA a different assay is needed and quantitative comparison across different miRNAs is more challenging. Moreover, RT-qPCR is less sensitive. Quantification of low expressed miRNAs may therefore be less accurate. As an example in cartilage, miR-329-3p was well detected with miRNAseq returning around 15 reads (BaseMean: 15.6 in 30 samples of the RAAK study [24]). When analyzing this miRNA in cartilage with RT-qPCR the Ct-values were around 30 using a miRNA-specific TaqMan assay, which is suboptimal for robust and reliable analyses (*see* **Note 8**).

5. To quantify well-expressed miRNAs, cDNA synthesis can be performed with TaqMan Advanced miRNA cDNA Synthesis Kit (Applied Biosystems) (*see* **Note 9**). The manual suggests using 1–10 ng of total RNA per reaction. We recommend to use 20 ng total RNA. Furthermore, follow the manufacturer's protocol.

4 Notes

1. Upon sampling of the cartilage for isolation of chondrocytes, be careful to avoid contamination with bone or synovium.

   Criteria for selection of preserved OA cartilage are as described in ref. 25: whiteness of the cartilage, surface integrity (avoid areas with visible fibrillation/crack formation), and both thickness as well as hardness of the cartilage upon sampling with a scalpel.
2. Cells can be maintained in culture with little dedifferentiation (or fast redifferentiation) up to passage 2–3 [27].
3. Counting cells using a Nucleocounter NC-200 (ChemoMetec) which applies state-of-the-art image cytometry ensures consistent cell counts over time. This additionally provides with percentage of cell viability and cell size.
4. Efficient isolation of high-quality miRNAs can be obtained with miRNeasy Kit with mini spin columns (Qiagen).

Table 3
Comparison BaseMean values miRNA-sequencing and Ct-values RT-qPCR for miRNAs expressed at different levels in articular cartilage

miRNA	BaseMean	Ct	Taqman assay
hsa-miR-329-3p	15.6	30.71	478029_mir
hsa-miR-99a-3p	17.9	27.83	479224_mir
hsa-miR-127-5p	52.9	23.57	477889_mir
hsa-miR-143-3p	11478.0	25.10	477912_mir
hsa-miR-451a	22519.5	18.05	478107_mir
hsa-let-7f-5p	35517.6	26.38	478578_mir
hsa-miR-26a-5p	38882.5	23.27	477995_mir

5. For long-term storage, RNA integrity is best maintained upon precipitation (add to 1× volume RNA: 2.5× volumes 100% EtOH and 0.1× volume 3 M NaAc; for very low RNA concentrations adding a carrier such as Glycogen will contribute to precipitation efficiency).
6. RIN values of at least 8 ensure high quality downstream analyses.
7. To obtain consistent results in downstream analyses for miRNAseq taking a threshold of 4 reads should suffice.
8. When using RT-qPCR, Ct-values depend on the TaqMan assay applied and as illustrated in Table 3 this may result in variable ratios between normalized read counts and Ct-values.
9. When analyzing lowly expressed miRNAs with RT-qPCR, the use of specific TaqMan assays instead of Advanced assays may deliver more sensitivity and thus more consistent results (Fig. 4). Note that use of specific TaqMan assays requires separate cDNA synthesis for each specific miRNA.

Acknowledgments

YFMR acknowledges the Foundation for Research in Rheumatology (FOREUM), BBMRI-NL Complementation project (CP2013-83), Ana Fonds (O2015-27), and Dutch Scientific Research Council NWO/ZonMW VICI scheme (nr 91816631/528) for financially supporting our work. The Leiden University Medical Center supports the RAAK study. We thank all study participants of the RAAK study. We are also grateful to all members of the Meulenbelt lab at the LUMC for valuable discussion, especially Dr. Coutinho de Almeida for expert suggestions with respect

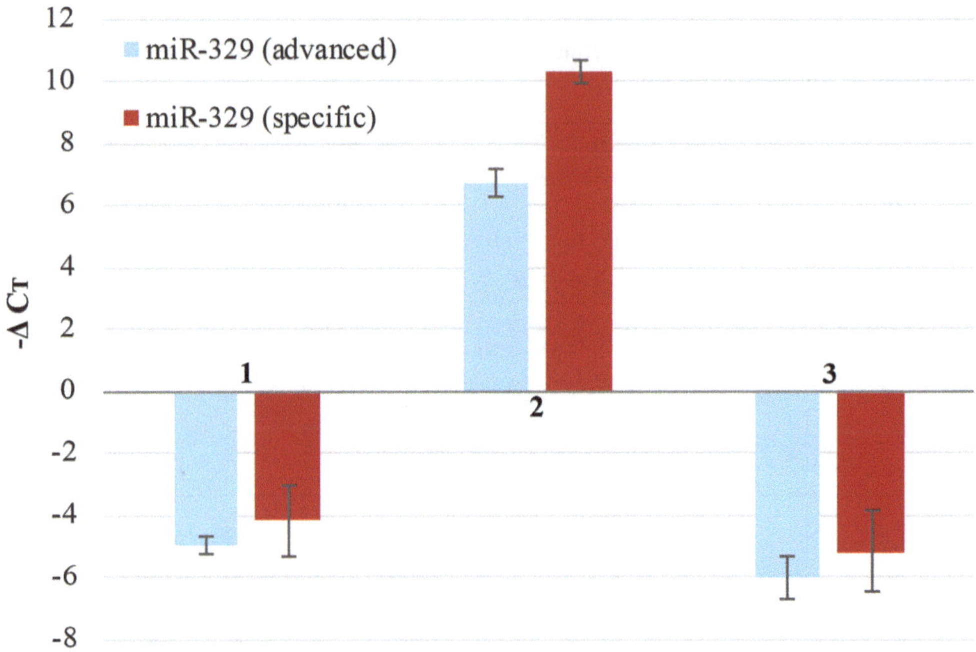

Fig. 4 RT-qPCR for miR-329-3p. Relative expression of miR-329-3p in proliferating primary chondrocytes as determined with specific (red) or with advanced (blue) TaqMan assays. Values show average -ΔCt-values ± SEM for 2 independent donors in triplicate. For calculations of the -ΔCt, U6 (specific) or miR-26a (advanced) were used as reference (legend: 1: control transfection; 2: transfection with miR-329-3p mimic; 3: transfection with miR-329-3p inhibitor)

to Next Generation Sequencing analysis. AM wishes to acknowledge financial support from The European Commission Seventh Framework Programme (EU FP7; HEALTH.2012.2.4.5-2, project number 305815 and Novel Diagnostics and Biomarkers for Early Identification of Chronic Inflammatory Joint Diseases). Details of the D-BOARD FP7 Consortium are available at: http://www.d-board.eu. The Innovative Medicines Initiative Joint Undertaking under grant agreement No. 115770, resources of which are composed of financial contribution from the European Union's Seventh Framework programme (FP7/2007-2013), EFPIA companies' in-kind contribution. Details of the APPROACH IMI Consortium are available at: https://www.approachproject.eu. Marie Curie Intra-European Fellowship for Career Development grant (project number 625746; acronym: C HONDRION; FP7-PEOPLE-2013-IEF). The European Structural and Social Funds through the Research Council of Lithuania (Lietuvos Mokslo Taryba) according to the activity "Improvement of researchers' qualification by implementing world-class R&D projects" of Measure No. 09.3.3-LMT-K-712 (grant application code: 09.3.3-LMT-K-712-01-0157, agreement No. DOTSUT-215). The European Structural and Social Funds through the

Research Council of Lithuania (Lietuvos Mokslo Taryba) according to the Programme "Attracting Foreign Researchers for Research Implementation," Grant No 0.2.2-LMTK-718-02-0022.
Author disclosures: YFMR has nothing to disclose. AM has consulted for the following companies in the last 3 years: Abbvie, Aché Laboratórios Farmacêuticos S.A., Galapagos, Kolon TissueGene, Pfizer Consumer Health (PCH), Servier, Bioiberica S.A., and Artialis S.A.

References

1. McKee TJ, Perlman G, Morris M, Komarova SV (2019) Extracellular matrix composition of connective tissues: a systematic review and meta-analysis. Sci Rep 9(1):10542
2. Ramos YF, Meulenbelt I (2017) The role of epigenetics in osteoarthritis: current perspective. Curr Opin Rheumatol 29(1):119–129
3. Coutinho de Almeida R, Ramos YFM, Meulenbelt I (2017) Involvement of epigenetics in osteoarthritis. Best Pract Res Clin Rheumatol 31(5):634–648
4. Kalamegam G, Memic A, Budd E, Abbas M, Mobasheri A (2018) A comprehensive review of stem cells for cartilage regeneration in osteoarthritis. Adv Exp Med Biol 1089:23–36
5. Swingler TE, Niu L, Smith P, Paddy P, Le L, Barter MJ, Young DA, Clark IM (2019) The function of microRNAs in cartilage and osteoarthritis. Clin Exp Rheumatol 37 Suppl 120 (5):40–47
6. Haseeb A, Makki MS, Khan NM, Ahmad I, Haqqi TM (2017) Deep sequencing and analyses of miRNAs, isomiRs and miRNA induced silencing complex (miRISC)-associated miRNome in primary human chondrocytes. Sci Rep 7(1):15178
7. Kobayashi T, Lu J, Cobb BS, Rodda SJ, McMahon AP, Schipani E, Merkenschlager M, Kronenberg HM (2008) Dicer-dependent pathways regulate chondrocyte proliferation and differentiation. Proc Natl Acad Sci U S A 105(6):1949–1954
8. Kobayashi T, Papaioannou G, Mirzamohammadi F, Kozhemyakina E, Zhang M, Blelloch R, Chong MW (2015) Early postnatal ablation of the microRNA-processing enzyme, Drosha, causes chondrocyte death and impairs the structural integrity of the articular cartilage. Osteoarthr Cartil 23 (7):1214–1220
9. Vinatier C, Dominguez E, Guicheux J, Carames B (2018) Role of the inflammation-autophagy-senescence integrative network in osteoarthritis. Front Physiol 9:706
10. Levine B, Kroemer G (2019) Biological functions of autophagy genes: a disease perspective. Cell 176(1–2):11–42
11. Rockel JS, Kapoor M (2016) Autophagy: controlling cell fate in rheumatic diseases. Nat Rev Rheumatol 12(9):517–531
12. Lotz MK, Carames B (2011) Autophagy and cartilage homeostasis mechanisms in joint health, aging and OA. Nat Rev Rheumatol 7 (10):579–587
13. Mobasheri A, Rayman MP, Gualillo O, Sellam J, van der Kraan P, Fearon U (2017) The role of metabolism in the pathogenesis of osteoarthritis. Nat Rev Rheumatol 13 (5):302–311
14. Pyo JO, Yoo SM, Ahn HH, Nah J, Hong SH, Kam TI, Jung S, Jung YK (2013) Overexpression of Atg5 in mice activates autophagy and extends lifespan. Nat Commun 4:2300
15. Fernandez AF, Sebti S, Wei Y, Zou Z, Shi M, McMillan KL, He C, Ting T, Liu Y, Chiang WC, Marciano DK, Schiattarella GG, Bhagat G, Moe OW, Hu MC, Levine B (2018) Disruption of the beclin 1-BCL2 autophagy regulatory complex promotes longevity in mice. Nature 558(7708):136–140
16. Zhang Y, Vasheghani F, Li YH, Blati M, Simeone K, Fahmi H, Lussier B, Roughley P, Lagares D, Pelletier JP, Martel-Pelletier J, Kapoor M (2015) Cartilage-specific deletion of mTOR upregulates autophagy and protects mice from osteoarthritis. Ann Rheum Dis 74 (7):1432–1440
17. Song J, Ahn C, Chun CH, Jin EJ (2014) A long non-coding RNA, GAS5, plays a critical role in the regulation of miR-21 during osteoarthritis. J Orthop Res 32(12):1628–1635
18. Wang WJ, Yang W, Ouyang ZH, Xue JB, Li XL, Zhang J, He WS, Chen WK, Yan YG, Wang C (2018) MiR-21 promotes ECM degradation through inhibiting autophagy via the PTEN/akt/mTOR signaling pathway in human degenerated NP cells. Biomed Pharmacother 99:725–734

19. Wu J, Kuang L, Chen C, Yang J, Zeng WN, Li T, Chen H, Huang S, Fu Z, Li J, Liu R, Ni Z, Chen L, Yang L (2019) miR-100-5p-abundant exosomes derived from infrapatellar fat pad MSCs protect articular cartilage and ameliorate gait abnormalities via inhibition of mTOR in osteoarthritis. Biomaterials 206:87–100
20. Cai C, Min S, Yan B, Liu W, Yang X, Li L, Wang T, Jin A (2019) MiR-27a promotes the autophagy and apoptosis of IL-1beta treated-articular chondrocytes in osteoarthritis through PI3K/AKT/mTOR signaling. Aging (Albany NY) 11(16):6371–6384
21. Endisha H, Rockel J, Jurisica I, Kapoor M (2018) The complex landscape of microRNAs in articular cartilage: biology, pathology, and therapeutic targets. JCI Insight 3(17): e121630
22. Deng Q, Hu H, Yu X, Liu S, Wang L, Chen W, Zhang C, Zeng Z, Cao Y, Xu-Monette ZY, Li L, Zhang M, Rosenfeld S, Bao S, Hsi E, Young KH, Lu Z, Li Y (2019) Tissue-specific microRNA expression alters cancer susceptibility conferred by a TP53 noncoding variant. Nat Commun 10(1):5061
23. Zhou Y, Wang X, Song M, He Z, Cui G, Peng G, Dieterich C, Antebi A, Jing N, Shen Y (2019) A secreted microRNA disrupts autophagy in distinct tissues of Caenorhabditis elegans upon ageing. Nat Commun 10(1):4827
24. Coutinho de Almeida R, Ramos YFM, Mahfouz A, den Hollander W, Lakenberg N, Houtman E, van Hoolwerff M, Suchiman HED, Rodriguez Ruiz A, Slagboom PE, Mei H, Kielbasa SM, Nelissen R, Reinders M, Meulenbelt I (2019) RNA sequencing data integration reveals an miRNA interactome of osteoarthritis cartilage. Ann Rheum Dis 78 (2):270–277
25. Ramos YF, den Hollander W, Bovee JV, Bomer N, van der Breggen R, Lakenberg N, Keurentjes JC, Goeman JJ, Slagboom PE, Nelissen RG, Bos SD, Meulenbelt I (2014) Genes involved in the osteoarthritis process identified through genome wide expression analysis in articular cartilage; the RAAK study. PLoS One 9(7):e103056
26. D'Adamo S, Alvarez-Garcia O, Muramatsu Y, Flamigni F, Lotz MK (2016) MicroRNA-155 suppresses autophagy in chondrocytes by modulating expression of autophagy proteins. Osteoarthr Cartil 24(6):1082–1091
27. Bomer N, den Hollander W, Suchiman H, Houtman E, Slieker RC, Heijmans BT, Slagboom PE, Nelissen RG, Ramos YF, Meulenbelt I (2016) Neo-cartilage engineered from primary chondrocytes is epigenetically similar to autologous cartilage, in contrast to using mesenchymal stem cells. Osteoarthr Cartil 24 (8):1423–1430

Chapter 14

Mitochondrial Biogenesis, Activity, and DNA Isolation in Chondrocytes

Yun Wang, Liang-Yu Chen, and Ru Liu-Bryan

Abstract

Chondrocytes, the only cells in articular cartilage, are metabolically active and responsible for the turnover of extracellular matrix and maintenance of the tissue homeostasis. Changes in chondrocyte function can cause degradation of the matrix and loss of articular cartilage integrity, leading to development and progression of osteoarthritis (OA). These changes are exemplified by accumulated mitochondrial damage and dysfunction. Because mitochondria are the critical organelles to produce energy and play a key role in cellular processes, the approaches to assess mitochondrial function under both physiological and pathological conditions enable us to uncover the mechanisms on how dysfunction of mitochondria in chondrocytes mediates signaling pathways that are involved in disturbance of cartilage homeostasis. In this chapter, we describe the methods to evaluate mitochondrial biogenesis, activity and mitochondrial DNA (mtDNA) integrity in chondrocytes.

Key words Osteoarthritis, Chondrocyte, Mitochondrial biogenesis, Mitochondrial DNA, Peroxisome proliferator-activated receptor γ coactivator 1α, Mitochondrial transcription factor A

1 Introduction

Osteoarthritis (OA) is the most common form of arthritis. The central feature of OA is the progressive degeneration of articular cartilage in the synovial joints, influenced by aging, biomechanical injury, metabolic disturbance, and other factors [1, 2]. Chondrocytes, the sole cell type in cartilage, regulate extracellular matrix turnover and maintain tissue homeostasis [2]. Mitochondria are critical subcellular organelles to produce energy and are involved in regulation of many cellular processes [3]. Human OA chondrocytes exhibit mitochondrial dysfunction, manifested by decreases in mitochondrial biogenesis including reduced mitochondrial DNA content, mitochondrial mass, mitochondrial respiratory complex expression and activity and intracellular ATP levels, increases in mitochondria-mediated oxidative stress, reduced antioxidant

Tariq M. Haqqi and Véronique Lefebvre (eds.), *Chondrocytes: Methods and Protocols*, Methods in Molecular Biology, vol. 2245, https://doi.org/10.1007/978-1-0716-1119-7_14,

capacity, and enhanced catabolic responses to inflammatory cytokines [4, 5].

Reduced mitochondrial biogenesis in OA chondrocytes is reflected by decreases in mitochondrial DNA content, mitochondrial mass, oxygen consumption, oxidative phosphorylation (OXPHOS), and intracellular ATP levels (Fig. 1). Decreased activity of AMP-activated protein kinase (AMPK) in OA chondrocytes is a core event compromising mitochondrial biogenesis, mediated through downregulation of the NAD^+-dependent protein deacetylase SIRT1 and peroxisome proliferator-activated receptor γ coactivator 1α (PGC-1α), a critical regulator of mitochondrial biogenesis [7]. PGC-1α induces expression of transcription of nuclear respiratory factors (NRF1 and NRF2) that promote mitochondrial transcription factor A (TFAM) gene expression [8], leading to an increase in transcription of mtDNA [9, 10]. Pharmacologic activation of AMPK can reverse established impairment of mitochondrial biogenesis in cultured advanced human knee OA chondrocytes via SIRT1-PGC-1α signaling [6].

Given that mitochondria are the main source of reactive oxygen species (ROS), mtDNA is vulnerable to oxidative damage, which can cause mtDNA mutations, promoting mitochondrial respiratory chain dysfunction and augmenting ROS production [11]. Accumulation of mtDNA damage is observed in human OA chondrocytes [12, 13]. The most frequent mtDNA mutation in human tissues is a 4977-bp deletion ($mtDNA^{4977}$), known as "the common deletion." This $mtDNA^{4977}$ deletion is prevalent in human knee OA chondrocytes, correlated with reduced expression and function of the mitochondrial-localized NAD-dependent deacetylase SIRT3 (Fig. 3). Deficiency of SIRT3 greatly increased acetylation of antioxidant enzyme manganese superoxide dismutase (MnSOD or SOD2) and the DNA repair enzyme 8-oxoguanine glycosylase (OGG) in chondrocytes, leading to decreased mitochondrial biogenesis and antioxidative oxidative stress and DNA repair capacity [13], therefore compromising mitochondrial function. Pharmacologic activation of the AMPK can eliminate preexisting mtDNA mutation in human OA chondrocytes and improved mitochondrial function via SIRT3-SOD2 and SIRT3-OGG1 signaling [13].

Here we describe the methods for studying mitochondrial biogenesis, activity and mtDNA integrity in chondrocytes. These methods have allowed us to further understand critical mitochondrial function in regulation chondrocyte homeostasis and study the potential of new therapeutic approaches (e.g., activation of AMPK) for improving mitochondrial function in OA chondrocytes.

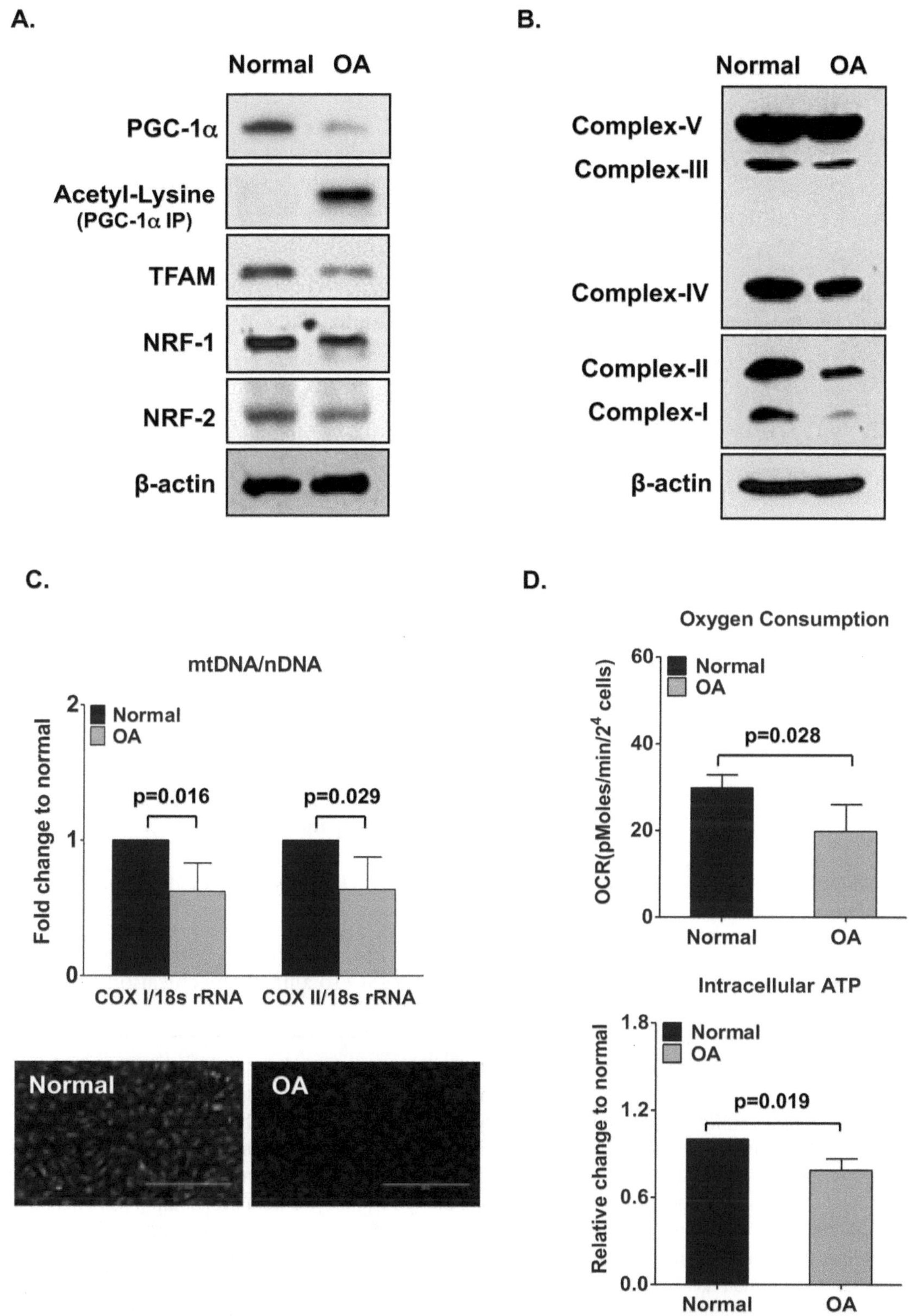

Fig. 1 Reduced mitochondrial biogenesis capacity and function in human osteoarthritis (OA) chondrocytes. (**a**) PGC-1α, acetyl-lysine PGC-1α, expression of mitochondrial transcriptional factor A (TFAM), nuclear respiratory factor 1 (NRF-1), and NRF-2 in normal donor and human OA chondrocyte lysates, as

2 Materials

Prepare all the solutions using deionized water or specifically indicated. Other common materials include scalpel, nylon cell strainer, DPBS, tubes, cell culture flasks and plates, incubators, centrifuges, and tissue culture hood.

2.1 Cell Culture Medium and Reagents

1. Dulbecco's Modified Eagle Medium (DMEM) containing 4.5 g/L glucose, L-glutamine, and without sodium pyruvate.
2. Fetal calf serum (FCS).
3. Streptomycin and penicillin (100×).
4. Type II collagenase (Worthington).
5. Trypsin (0.5 mg/ml).
6. Chondrocyte complete media: add 100 ml fetal calf serum (FCS) and 10 ml Pen/strep (100×) to 890 ml DMEM. The final complete media for chondrocyte culture is DMEM with 10% FCS, 100 μg/ml streptomycin, and 100 IU/ml penicillin.
7. Chondrocyte digestion media: 2 mg/ml Type II collagenase in 5% FCS DMEM.
8. Primary human knee chondrocytes are obtained from autopsy donors (*see* **Note 1**). The autopsy donors are graded macroscopically according to a modified Outer bridge scale [14]. Grade I represents intact cartilage surface (normal); grade II represents minimal fibrillation (OA); grades III and IV represent overt fibrillation and erosion (OA).

2.2 Mitochondrial Mass Determination

1. Fluorescent probe MitoTracker Green FM (*see* **Note 2**).
2. DMSO.
3. Prepare stock solutions: dissolve the lyophilized MitoTracker Green in high-grade DMSO to a final concentration of 1 mM.

Fig. 1 (continued) determined by Western blotting. (**b**) Expression of mitochondrial respiratory complexes I–IV and ATP synthase (complex V) in normal and OA chondrocytes, as determined by Western blotting. (**c**) Top, Mitochondrial DNA (mtDNA) content as determined by the mtDNA-encoded cytochrome c oxidase subunit I or II (*COX1* or *COX2*): nucleus-encoded 18S ribosomal DNA (18S rDNA) ratio. Bottom, Mitochondrial mass, as determined by MitoTracker Green FM staining. Values are the mean ± SD and are from a representative experiment of five independent experiments performed using chondrocytes from five different pairs of normal and OA (grade III or IV) donors. Original magnification × 10. (**d**) Oxygen consumption rate (OCR) (top) and intracellular ATP level (bottom) in normal and OA chondrocytes. Values are the mean ± SD of 3 independent experiments performed using chondrocytes from 3 different pairs of normal and OA (grade III or IV) donors. Results in **a** and **c** are representative of experiments using chondrocytes from 5 different pairs of normal and OA (grade III or IV) donors. nDNA = nuclear DNA. (This figure is reproduced from ref. 6 with permission from John Wiley & Sons)

4. Prepare working solution: dilute 1 mM MitoTracker stock solution to the final work concentration 20 nM–200 nM in growth medium.
5. Fluorescence microscope with Ex 490 nM and Em 516 nM.

2.3 Mitochondrial DNA Measurement

1. DNeasy Blood & Tissue kit (*see* **Note 3**).
2. Ethanol (96–100%).
3. Prepare buffer AW1 and buffer AW2, add the appropriate volume of ethanol (96–100%) as indicated on the bottle and shake well. The buffer is stable for at least 1 year after adding ethanol at room temperature.
4. Real-time PCR machine (e.g., LightCycler 480 system [Roche] 96-well).
5. NanoDrop spectrophotometer.
6. DNA primers.

   18 s rRNA forward primer: 5′-TAGAGGGACAAGTGGCG TTC-3′

   18 s rRNA reverse primer: 5′- CGCTGAGCCAGTCAGTGT-3′

   mtCOXI forward primer: 5′-GGCCTGACTGGCATTGTA TT- 3′,

   mtCOXI reverse primer: 5′-TGGCGTAGGTTTGGTCTA GG-3′,

   mtCOXII forward primer: 5′-GCCGACTAAATCAAGCAA CA-3′,

   mtCOXII reverse primer: 5′-CAATGGGCATAAAGCTA TGG-3′,
7. SYBR Green I Master (Roche) (*see* **Note 4**).
8. Prepare PCR mix of SYBR Green I Master and primers according to 20 μl standard reaction volume per well: 10 μl SYBR Green Master (2×), 1 μl forward primer (10 μM stock), 1 μl reverse primer (10 μM stock), and 3 μl PCR-grade H_2O.

2.4 Mitochondrial DNA Mutation Detection

1. Platinum *Taq* DNA polymerase high fidelity.
2. Primers: for detection of deletion mtDNA^{4977} mutant: L8150L + H13650, WT: L3304 + H3836).

   L8150L: 5′ CCGGGGGTATACTACGGTCA 3′.

   H13650: 5′ GGGGAAGCGAGGTTGACCTG 3′.

   L3304: 5′ AACATACCCATGGCCAACCT 3′.

   H3836: 5′ GGCAGGAGTAATCAGAGGTG 3′.

3. PCR reaction setup: 5 μl 10× high Fidelity PCR buffer, 2 μl 50 mM $MgSO_4$, 1 μl 10 mM dNTP mix, 1 μl forward primer, 1 μl reverse primer, 1 μl DNA, 0.2 μl *Taq*, 38.8 μl nuclease free water.

2.5 SDS/Western Blot Reagents

1. RIPA buffer: 150 mM NaCl, 5 mM EDTA pH 8.0, 50 mM Tris, pH 8.0, NP-40, 1%, 0.5% sodium deoxycholate, 0.1% SDS, 2 mM sodium vanadate and 1 × protease inhibitor cocktail.
2. Pierce BCA protein assay

   Albumin (BSA) standards dilution: prepare a series of BSA standard stock solution and conduct a series of two-fold serial dilutions to generate 0.0625, 0.125, 0.25, 0.5, 1 and 2 mg/ml protein standards.

   BCA Working Reagent: mix 50 parts of BCA reagent A with 1 part of BCA reagent B (Reagent A: B, 50:1).
3. Antibodies

   Primary antibody: anti-PGC-1α, anti-TFAM, anti-NRF1, anti-NRF2, anti-OXPHOS, anti-SOD2, anti-acetylated SOD2, anti-OGG1, anti-acetylated OGG1, anti-acetyl lysine, and anti-β-Actin. Secondary antibody: donkey anti-mouse IgG (H + L) HRP, donkey anti-rabbit IgG (H + L) HRP.
4. SuperSignal West Pico Chemiluminescent Substrate (Thermo Fisher Scientific): working solution made by mixing equal parts of detection reagents 1 and 2 in the kit. Use 0.125 ml working solution per cm^2 of membrane.
5. 4–20% SDS-PAGE gradient gel (*see* **Note 5**)
6. SDS-PAGE running buffer: 0.025 M Tris–HCI, PH8.3, 0.192 M glycine, 0.1% SDS (*see* **Note 6**).
7. Western blot transfer buffer: 0.025 M Tris–HCI, 0.192 M glycine, 20% methanol (*see* **Note 6**).
8. Western blot wash buffer (TBST): 20 mM Tris–HCI and 150 mM NaCl, pH 7.4 containing 0.05% Tween-20 (*see* **Note 6**).
9. PVDF transfer membranes.
10. Blocking/antibody diluent solution: 5% milk or 5% BSA in TBST.

2.6 Assessment of Acetylation Levels of PGC-1a

1. Pierce classic IP kit for PGC-1α immunoprecipitation (Thermo Fisher Scientific).
2. Primary antibody anti-PGC-1α for IP (Santa Cruz Biotechnology), anti-acetyl lysine antibody (Cell Signaling).

2.7 siRNA Transfection Reagent

1. X-tremeGene siRNA transfection reagent (*see* **Note 7**).
2. TFAM siRNA and control siRNA (Thermo Fisher Scientific).
3. Opti-MEM® I Reduced-Serum Medium.

2.8 SIRT3 Deacetylation Activity Assay

1. SIRT3 activity assay kit (Abcam).
2. 96-well plate—black wall
3. Microplate reader capable of measuring fluorescence at Ex/Em = 340–360/440–460 nm.
4. Orbital shaker.

2.9 Oxygen Consumption Rate (OCR) Measurement

1. Agilent Seahorse XFe/XF analyzers.
2. XFe 96 FluxPax (containing sensor cartridge, cell culture microplates, and calibrant solution. Agilent Tech).
3. XF 1 M Glucose solution (Agilent Tech).
4. XF DMEM medium (Agilent Tech). The glucose is added into Agilent DMEM to make final concentration of 4500 mg/l, since the basal medium does not contain glucose. Adjust pH to 7.4.
5. Seahorse XF cell Mito Stress test kit (Agilent Tech).
6. Prepare oligomycin, FCCP, and rotenone/antimycin A into the final concentration of 2.5 μM, 1 μM, and 0.5 μM per well (optimal concentrations for human OA chondrocytes).

2.10 Intracellular ATP Measurement

1. ATP colorimetric/fluorometric assay kit (Biovision) (*see* **Note 8**).
   (a) Buffer preparation: warm ATP assay buffer to room temperature before use.
   (b) ATP converter: dissolve ATP converter in 220 μl ATP assay buffer. Aliquot and store at −20 °C.
   (c) Developer Mix: Dissolve in 220 μl ATP Assay Buffer. Aliquot and store at −20 °C.
   (d) ATP Standard: Dissolve in 100 μl of distilled water to generate 10 mM stock solution. Keep on ice while in use. Store at −20 °C.
   (e) Sample preparation: Lyse 1 × 10^6 cells in 100 μl ATP Assay Buffer and deproteinize using the Kit (BioVision) or 10 kDa Spin Column (Millipore).
   (f) Standard Curve Preparation: dilute 10 μl of the ATP Standard with 90 μl of dH_2O to generate 1 mM ATP standard, mix well.
2. 96-well plate with flat bottom
3. Multi-well spectrophotometer (OD 570 nM).

3 Methods

3.1 Primary Human Knee Chondrocyte Cell Preparation and Culture as Reported [15, 16]

1. Human knee chondrocytes are isolated from autopsy donors that are graded as described in Subheading 2.1, **item 8**.
2. Before the chondrocyte isolation, the cartilage surfaces of autopsy donor are rinsed with saline first and cartilage sections are cut at around 5 mm apart by scalpel vertically from the cartilage surface onto the subchondral bone. These strips are washed with DMEM (*see* **Note 9**).
3. After discarding the supernatant wash, the cartilage slices are cut into 3 mm × 3 mm pieces and digested with double volume of digestion media (from Subheading 2.1, **item 7**) on a gyratory shaker at 37 °C overnight.
4. The next day, the digested mixture is filtered through 100 μm nylon cell strainer and human chondrocytes are pelleted by centrifuge and washed using complete DMEM. And then the cells are resuspended in compete DMEM media and seeded into the culture flasks (3×10^6 cells/T75 flask). The cells are incubated at 37 °C in a humidified gas mixture containing 5% CO_2 balanced with air and ready for use upon confluency (*see* **Note 10**).
5. Unless indicated, chondrocytes are plated at 3×10^5 cells per well in 12 well plates or 7.5×10^5 cells per well in 6-well plates.

3.2 Protein Expression of PGC-1α, TFAM, NRF1, NRF2, SIRT3, SOD2, OGG1, Acetylated SOD2, Acetylated OGG1, and OXPHOS

3.2.1 Cell Lysate Preparation

1. Human chondrocyte cells are harvested after desired treatment and centrifuged at 500 × *g*, 4 °C for 5 min, and are washed once with PBS.
2. The cell pellets are resuspended in RIPA buffer with 2 mM sodium vanadate and protease inhibitor cocktail (1×10^6 cells in 100 μl) and sonicated for 5 s, and then lysed for half an hour on ice.
3. The mixture of lysed cells is centrifuged at 12,000 × *g*, 4 °C for 10 min and the supernatant is the final protein extract.

3.2.2 Protein Concentration Determined by BCA

1. Dilute the protein supernatant from Subheading 3.2.1 if needed.
2. BCA Working Reagent is mixed as described in Subheading 2.5, **item 2**.
3. Microplate procedure: pipet 5 μl standards or samples into 100 μl BCA working solution and mix the plate thoroughly.
4. Cover the plate and incubate for 30 min at RT and measure the absorbance at 562 nM on a plate reader. Determine the protein concentration for each sample based on the BSA standard curve.

5. Prepare protein extracts of cell lysates with loading buffer and heat at 95 °C for 5 min.

3.2.3 Protein Separation and Membrane Transfer

1. Centrifuge the heated samples at 3000 × *g* for 30 s to bring down the condensate.
2. Load the samples (15–20 μg) to 4–20% SDS-PAGE gradient gels, and electrophoreses them at constant 80 volts (V) until the samples enter the gel and then continue at 150 V till the dye reaches the bottom of the gel.
3. Prepare the Polyvinylidene difluoride (PVDF) membrane and quickly immerse in methanol for 30 s, after rinsing in water, put the membrane in transfer buffer until use.
4. Following electrophoresis, prepare the wet-sandwich transfer cassette and run at constant current 250 mA for 90 min in cold room or ice box; or use semi-dry transfer at 15 V for 25 min.

3.2.4 Western Blotting

1. After transfer, rinse the membrane with TBST and incubate in blocking solution for 30 min to 1 h at RT.
2. Incubate the membrane with appropriate primary antibodies (OXPHOS diluted at 1:300 in 5% milk. Others are diluted at 1:1000 in 5% BSA) on a shaker at 4 °C overnight.
3. The next day, wash the membrane thoroughly with TBST 3 times (5 min each time), and then add secondary antibody (1:3000 in 5% milk) for 1 h at RT.
4. Wash the membrane 3 times with TBST and incubate the membrane with working solution from Subheading 2.5, **item 4** for 1 min at room temperature.
5. Place the membrane in a plastic sheet protector and place it in a film cassette. Place the film on top of the membrane. Vary the exposure time to achieve optimal results. Figures 1a, b and 3b are examples of Western blot showing that PGC-1α, NRF-1, NRF-2, OXPHOS, and SIRT3 decreased in OA chondrocytes.

3.3 Assessment of Acetylation Levels of PGC-1α

3.3.1 Immunoprecipitation with PGC-1α Antibody Using Pierce Classic IP Kit (See ***Note 11****)*

1. Human chondrocytes are cultured in 6-well plate with and without treatment of testing agents for desired time. After removing the culture media, the cells are washed once with cold DPBS.
2. Add 400 μl ice cold IP lysis/wash buffer into the cells and incubate on ice for 5 min with periodic mixing.
3. Transfer the lysate to a microcentrifuge tube and incubate on ice for another 20 min, then centrifuge at ~13,000 × *g* for 10 min to pellet the cell debris.
4. Transfer supernatant to a new tube and determine protein concentration as described in Subheading 3.2. The cell lysate is ready for preclear by control agarose resin.

5. Prepare enough control agarose resin slurry (80 μl for 1 mg lysate), add the slurry into a spin column and wash once with PBS, centrifuge, and discard the flow-through.
6. Add 1 mg lysate to the column containing the resin and incubate at 4 °C for 1 h with gentle end-over-end mixing.
7. Centrifuge column at 1000 × *g* for 1 min. Discard the column containing the resin and save the flow-through, which will be cell lysate for next step.
8. Determine the concentration of cell lysate from **step 7** as described in Subheading 3.2.2.
9. Add 5 μg PGC-1α primary antibody into 500 μg cell lysate in a microcentrifuge tube and dilute the antibody–lysate solution to 300–600 μl with IP Lysis/Wash Buffer. Incubate with gentle rocking overnight at 4 °C to form antibody–sample complex.
10. Gently swirl the bottle of Pierce Protein A/G Agarose to obtain an even suspension. Using a cut pipette tip, add 50 μl of the resin slurry into a Pierce Spin Column. Place column into a microcentrifuge tube and centrifuge at 1000 × *g* for 1 min. Discard the flow-through.
11. Wash resin twice with 100 μL of cold IP Lysis/Wash Buffer. Discard the flow-through after each wash.
12. Gently tap the bottom of the spin column on a paper towel to remove excess liquid and insert the bottom plug.
13. Add the antibody–sample complex to Protein A/G plus Agarose in the spin column. Attach the screw cap and incubate column with gentle end-over-end mixing or shaking for 1–2 h.
14. Remove bottom plug, loosen the screw cap, and place the column in a collection tube. Centrifuge column and save the flow-through. Do not discard flow-through until confirming that the IP has been successful.
15. Remove the screw cap, place the column into a new collection tube, add 200 μl of IP Lysis/Wash Buffer, and centrifuge.
16. Wash the resin three times with 200 μl IP Lysis/Wash Buffer and centrifuge after each wash.
17. Wash the resin once with 100 μl of 1× Conditioning Buffer (diluted from 100× with ultrapure water).
18. Sample-buffer elution: place the spin column containing the resin into a new collection tube and add 50 μl of 2× loading buffer. Keep the column unplugged and in the collection tube, and incubate at 100 °C for 5–10 min. Centrifuge to collect eluate. Allow the sample to cool to room temperature before applying to the SDS-PAGE gel.

3.3.2 Western Blot with Anti-Acetyl Lysine Antibody

Western blot is performed with anti-acetyl-lysine as described in Subheading 3.2. As depicted in Fig. 1a, acetylation level of PGC-1α increases in OA chondrocytes.

3.4 Mitochondrial Mass by MitoTracker Green FM Staining

1. Before staining, all the reagents are warmed up to room temperature and working solution is prepared as described in Subheading 2.2 (*see* **Note 12**).
2. Human chondrocytes are seeded into 6 well plates with the complete medium and treated with appropriate compound for desired time.
3. When the cells reach the desired confluency and treatment, replace the growth medium with prewarmed (37 °C) staining solution (*see* **Note 13**) and place the plates back in an incubator (*see* **Note 14**).
4. After incubation for 30 min, staining solution is replaced with the prewarmed media, and fluorescence stained cells are visualized under a fluorescence microscope (*see* **Note 15**). The example shown in Fig. 1c depicts that mitochondrial mass is much less in OA compared to normal chondrocytes.

3.5 mtDNA Content by Real-Time PCR

3.5.1 Total DNA Is Isolated from Chondrocytes Using DNeasy Blood & Tissue Kit

1. Buffer AW1 and buffer AW2 are prepared as described in Subheading 2.3.
2. Centrifuge the appropriate number of cells (maximum 5×10^6) for 5 min at 300 × *g*. Resuspend the pellet in 200 μl PBS, add 20 μl proteinase K.
3. Add 200 μl Buffer AL (without added ethanol), mix thoroughly by vortexing for 5–10 s, and incubate at 56 °C for 10 min.
4. Add 200 μl ethanol (100% ethanol) to the sample and mix thoroughly by vortexing to generate a homogeneous solution.
5. Pipet the mixture from last step into the DNeasy Mini spin column placed in a 2 ml collection tube. Centrifuge at 6000 × *g* for 1 min. Discard the flow-through and collection tube.
6. Place the column in a new 2 ml collection tube, add 500 μl buffer AW1, centrifuge at 6000 × g for 1 min. Discard the flow-through and collection tube.
7. Place the column in a new 2 ml collection tube, add 500 μl buffer AW2, centrifuge at 20,000 × *g* for 3 min. Discard the flow-through and collection tube (*see* **Note 16**).
8. Place the column in a clean 1.5 ml or 2 ml Eppendorf tube and pipet 200 μl Buffer AE onto dry DNeasy membrane. Incubate at RT for 1 min, then centrifuge for 1 min at 6000 × *g* to elute.

3.5.2 DNA Concentration Measured by NanoDrop

1. Turn on the computer attached to the NanoDrop.
2. Wash the NanoDrop pedestal using purified water.
3. Open the NanoDrop software and click on the 'Nucleic Acid' button. Add water to the pedestal and wait for the machine to initialize. When it's done, life the upper arm and dry the pedestal with a wipe.
4. Add 2 μl water or elution buffer to blank the NanoDrop. Dry the pedestal with wipe once done (*see* **Note 17**).
5. Add 2 μl sample from Subheading 3.5.1 to the pedestal and click the 'measure' button for measurement, and then export the data to excel.
6. Clean the pedestal when finishing all the measurement.

3.5.3 Real-Time PCR

1. Pipet 15 μl PCR mix from Subheading 2.3, **item** 7 into each well.
2. Add 5 μl DNA template Mitochondrial DNA (0.1 ng) or nuclear DNA (10 ng) into each well. Seal the plates and centrifuge at 1500 × *g* for 2 min to bring the mixture to the bottom.
3. Load the plate into the Real-time PCR machine (e.g., Light-Cycler 480 instrument) and start the quantitative real-time PCR program: initial denaturation at 94 °C for 10 min, followed by 40 cycles of denaturation at 95 °C for 15 s, annealing/extension and fluorescence reading at 60 °C for 1 min.
4. Fold changes of mtDNA-encoded cytochrome c oxidase subunit I (COXI) and COXII are calculated by 2-ΔΔct methods. 18 s rDNA is used as internal standard. Ratio of COXI or COXII DNA copies to 18S rDNA represents relative mitochondrial copy number. The examples shown in Figs. 1c and 2b demonstrate that mtDNA COXI and COXII are significantly decreased in OA chondrocytes.

3.6 OCR Measurement

1. Prepare XF medium (from Subheading **2.9**, **item 3**).
2. Plate 5 × 10^3 cells into each well/Agilent 96-well cell culture microplate with regular culture medium (*see* **Note 18**).
3. Incubate cells in the 5% CO_2 incubator overnight to allow the cells to adhere to the plate.
4. Add 200 μl sterile water into the utility plate, immerse sensor into water, and place the module into 37 °C oven (non-CO_2) overnight (*see* **Note 19**).
5. On the day of experiment, aspirate water and replenish with prewarm XF calibration solution for at least 1 h.
6. Wash the cells with XF medium twice to remove buffered medium. Add 180 μl of XF medium into each well and place plate in 37 °C oven (non-CO_2) for at least 45 min to 1 h prior to the assay.

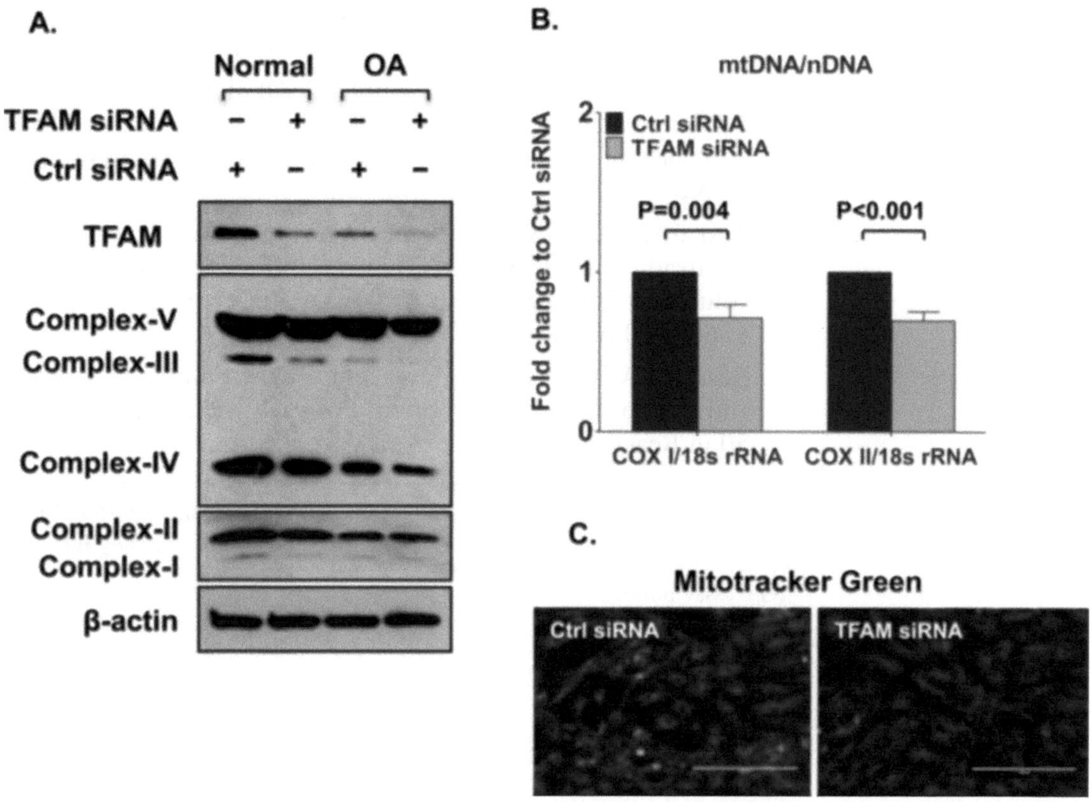

Fig. 2 Decreased mitochondrial biogenesis with knockdown of mitochondrial transcriptional factor A (TFAM). (**a**) Expression of mitochondrial respiratory complexes I–IV and ATP synthase (complex V) in normal and OA chondrocytes transfected with TFAM siRNA or control siRNA, as determined by Western blotting. Results are representative of 5 independent experiments using five different pairs of normal donors and donors with OA. (**b**) and (**c**) Mitochondrial DNA (mtDNA) content (**b**) and mitochondrial mass (**c**) in human knee chondrocytes (passage 1) from both normal donors and donors with osteoarthritis (OA) (grade IV) transfected with TFAM siRNA and control (Ctrl) siRNA, as determined by the mtDNA-encoded cytochrome c oxidase subunit I or II (*COX1* or *COX2*): nucleus-encoded 18S ribosomal DNA (18S rDNA) ratio. Values are the mean ± SD and are representative of three independent experiments performed using chondrocytes from three different donors. (This figure is reproduced from ref. 6 with permission from John Wiley & Sons)

7. Load 20 μl of 2.5 μM oligomycin (from Subheading 2.9, **item 6**) into "A" position of cartridge, 22 μl of 1 μM FCCP into "B" position of cartridge, and 25 μl of 0.5 μM rotenone/antimycin A into port "C" position of cartridge.
8. Place the cartridge with calibration fluid plate into instrument tray of Seahorse analyzer and click "Continue." It takes 15–30 min to complete.
9. When prompted, replace the calibration plate with the cell culture plate then click start. The example in Fig. 1d showsthat OA chondrocytes have significantly lower oxygen consumption rate.

3.7 ATP Measurement

1. Assay buffer, ATP converter, developer mix, ATP standard, and sample are prepared as described in Subheading 2.10.
2. Add appropriate volume of sample (2–50 μl; from Subheading 2.10, **item 1**(**e**)) to a 96-well plate. Adjust the volume to 50 μl/well with ATP Assay Buffer (from Subheading 2.10, **item 1**(**a**); *see* **Note 20**).
3. Add 0, 2, 4, 6, 8, and 10 μl of ATP Standard solution (*see* Subheading 2.10, **item 1(f)**) into a series of wells and adjust volume to 50 μl/well with ATP Assay Buffer to generate 0, 2, 4, 6, 8, 10 nmol/well of ATP Standard.
4. Reaction Mix: Mix enough reagent for the number of samples and standards to be performed: For each well, prepare a total 50 μl Reaction Mix:

	Colorimetric assay (μl)	Fluorometric assay (μl)
ATP assay buffer	44	45.8
ATP probe	2	0.2
ATP converter (from Subheading 2.10, **item 1(b)**)	2	2
Developer (from Subheading 2.10, **item 1(c)**)	2	2

   Mix well. Add 50 μl of the Reaction Mix to each well containing the ATP Standard and test samples.

5. Measurement: Incubate at room temperature for 30 min, protected from light. Measure absorbance (OD 570 nm) or fluorescence (Ex/Em = 535/587 nm) in a microplate reader. The signals are stable for 2 h.
6. Calculation: Correct background by subtracting the value derived from the 0 ATP standards from all readings. If the background control reading is significant, subtract the background control reading from sample reading. Plot the Standard Curve. Apply ATP sample readings to the standard curve to get B nmol of ATP in the sample well.

$$\text{Sample ATP concentration (C)} = B/V \times D$$
$$= \text{nmol}/\mu\text{l or } \mu\text{mol/ml or mM}$$

where: B is ATP amount in the reaction well from standard curve (nmol).

V is the sample volume added into the sample well (μl).

D is the dilution factor.

The example shown in Fig. 1d illustrates that OA chondrocytes contain lower levels of ATP compared to normal healthy chondrocytes (Fig. 1d).

3.8 Knockdown of TFAM in Human Knee Articular Chondrocytes

3.8.1 Cell Preparation Before Transfection

Cells are seeded in 6-well plate (2 ml media/well) the day before and reach 70% confluence on transfection day.

3.8.2 Transfection (Single Well of 6-Well Plate as an Example)

1. Dilute 10 μl X-tremeGENE siRNA transfection reagent with 100 μl serum/antibiotics-free Opti-MEM I Medium.
2. Dilute 2 μg siRNA with 100 μl serum/antibiotics-free Opti-MEM I Medium.
3. Mix the transfection reagent and siRNA complex by pipetting up and down and incubate for 15–20 min at RT (*see* **Note 21**).
4. Add the complex directly to the cells and incubate until assay for gene knockdown is performed (72 h) (*see* **Note 22**).
5. Levels of TFAM expression is examined by SDS-PAGE/Western blot analysis. Figure 2 demonstrates that knocking down TFAM expression (A) results in lower OXPHOS that is indicated by reduced expression subunits of mitochondrial respiratory complexes (A), and reduced mitochondrial mass (Fig. 2c) and DNA content (Fig. 2b).

3.9 mtDNA Mutation Detection by Real-Time PCR

1. Total DNA isolation from chondrocytes and DNA concentration measurement are as described as 3.5.
2. The PCR reaction (from Subheading 2.4, **item 3**) is carried out at 94 °C for 15 s, 58 °C for 15 s, and 72 °C for 40 s per cycle for total 35 cycles.
3. Products are separated by electrophoresis on a 1.5% agarose gel. The example in Fig. 3a shows that OA chondrocytes have a higher occurrence rate for $mtDNA^{4977}$ deletion mutation.

3.10 SIRT3 Deacetylase Activity Assay (See Note 23)

1. 5×10^5 chondrocytes are cultured in 6-well plate with and without treatment of testing agents for desired time. After removing the culture media, the cells are washed once with cold DPBS. The cells are collected in 500 μl cold DPBS using a cell scraper and are then transferred into 1.5 ml Eppendorf tube.
2. After centrifugation at 200 × *g* for 10 min at 4 °C, the cell pellets are resuspended with fresh cold 30 μl DPBS and are dispersed using a micro tube homogenizer for 5–10 s on ice.
3. The cell lysates are centrifuged at 12500 × *g* for 10 min at 4 °C. The supernatant is collected for SIRT3 activity assay.

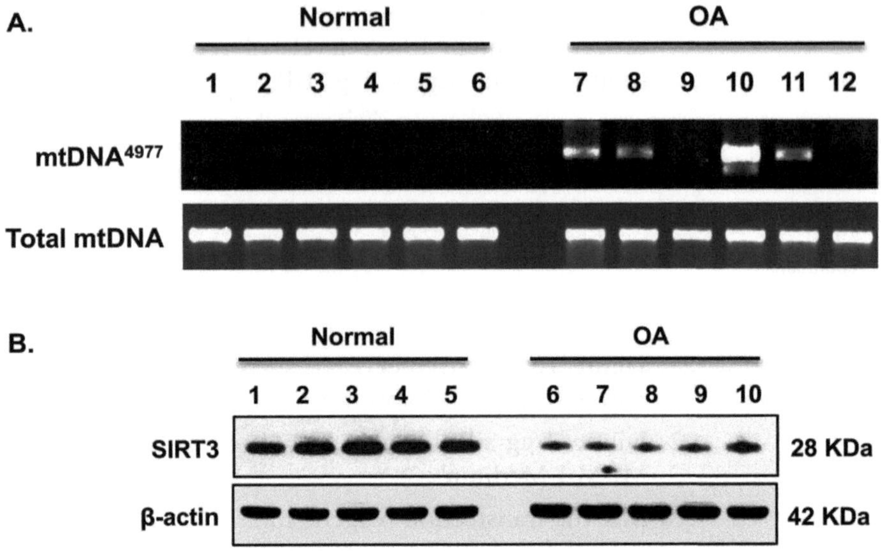

Fig. 3 Occurrence of mtDNA^{4977} deletion mutation in chondrocytes associated with SIRT3 deficiency. (**a**) Total DNA was directly isolated from human knee cartilage from normal and OA donors. PCR analyzed mtDNA^{4977} deletion mutation and total mtDNA as described in the Methods. Data in A represent primary chondrocytes from six normal and six OA donors. (**b**) Cell lysate SIRT3 expression was examined by Western blot analysis. Data in B represent five normal and five OA donors. (This figure is reproduced from ref. 13 with permission from Elsevier)

4. The fluorometric SIRT3 activity assay kit is used to measures the enzymatic activity of SIRT3 by the basic principle of changing a SIRT3 reaction into the activity of the peptidase. In this kit, fluorophore and quencher are coupled to amino terminal and carboxyl terminal of substrate peptide, respectively, and before reaction of deacetylase, the fluorescence cannot be emitted. However, if SIRT3 performs deacetylation, substrate peptide will be cut by the action of peptidase added simultaneously, quencher will separate from fluorophore, and fluorescence will be emitted. Deacetylase enzyme activity is measured by measuring this fluorescence intensity.
5. Follow the manufacturer's assay protocol exactly to quantify SIRT3 activity (*see* **Note 24**). The example illustrated in Fig. 4 shows that SIRT3 activity decreases in OA chondrocytes.

4 Notes

1. Human chondrocytes isolated from deidentified human knee donors are performed in compliance with institutional IRB reviewed and approved human subjects' protocols.
2. The kit is stable for at least 6 months if stored in ≤–20 °C upon arrival and protect from light.

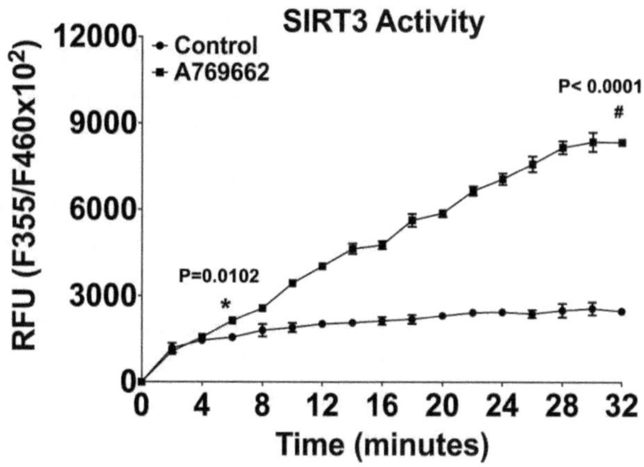

Fig. 4 Pharmacologic activation of AMPK significantly improved SIRT3 activity in OA chondrocytes. Cultured primary human knee OA chondrocytes were treated with AMPK pharmacological activator A-769662 (0.125 mM) for 2 h, and the SIRT3 activity was measured. (This figure is reproduced from ref. 13 with permission from Elsevier)

3. All the buffers in this kit should be stored dry at room temperature (15–25 °C) and are stable for 1 year. Other equipment or reagent needed: Vortexer, ethanol (96–100%), microcentrifuge tubes (1.5 ml or 2 ml), thermomixer, shaking water bath, or rocking platform for heating at 56 °C, PBS, pH 7.2 (50 mM potassium phosphate, 150 mM NaCl).
4. Keep the master away from light and store the kit at −15 to −25 °C until expiration date.
5. 4–20% Mini-PROTEAN® TGX™ Precast Protein Gels are bought from Bio-Rad or self-made in house.
6. 10 × stock of the buffers are prepared and dilute into 1× using DI H_2O on use day. Do not reuse all these 1× buffers.
7. X-tremeGene siRNA transfection reagent is stored at 2–8 °C. Do not freeze. Serum-free culture medium without supplements or Opti-MEM I medium is recommended for diluting the siRNA or the reagent.
8. The entire kit is protected from light and stored at −20 °C. ATP probe, ATP converter and developer mix are dissolved and aliquoted to store in −20 °C. They should be used within 2 months.
9. Use best aseptic practice to keep sample sterile.
10. Media is replaced every 3–4 days until use, and it usually takes 2–3 weeks for the chondrocytes to reach confluency in primary culture. Only passage 1 chondrocytes are used in all the experiments.

11. Perform all steps at 4 °C, unless otherwise indicated and all the resin centrifugation steps for 30–60 s at low speed (i.e., 1000 × *g*). A lysate preclearing step to reduce nonspecific binding to Protein A/G agarose beads is optional.
12. Allow the product to warm to room temperature. After dissolving the vial, aliquot the stock solution if needed and store them at ≤ −20 °C and protect from light, avoid multiple rethaw.
13. Optimize the dye concentration and try to keep it as low as possible. It will cause potential artifacts or toxicity at higher dose.
14. The incubation time varies and needs to be optimized. 15–45 min incubation is generally sufficient.
15. The cells are visualized right after staining, but the stained cells can be fixed for later analysis. Briefly, the cells are washed with prewarmed growth medium and then fixed with 2–4% formaldehyde growth medium for 15 min at 37 °C. After fixation, the cells are rinsed several times and kept in the medium until use.
16. At the last wash step, remove the column carefully to avoid any contamination of column from the flow-through ethanol. If it occurs, just repeat the centrifugation again in a new collection tube.
17. Use the same buffer as samples to blank the instrument. For example, if the sample is eluted in water, then use water for blank.
18. This is the optimal cell density to perform this experiment in our hand. It needs to optimize the cell density before performing this experiment.
19. Seal the sensor in the same container to prevent dry out and keep sterile.
20. Use fresh sample or snap frozen sample. Optimize the sample amount or dilution to make sure the reading is in the standard curve range.
21. Diluted siRNA should be combined with diluted transfection reagent within 5 min.
22. Removal of growth medium is not necessary.
23. All procedures are performed on ice or at 4 °C.
24. The reaction is initiated and the fluorescence intensity is measured by mixing simultaneously fluorescence-labeled acetylated peptide (the substrate), SIRT3, NAD and the developer. Since the reaction is not stopped, it is necessary to measure fluorescence intensity at regular intervals after the reaction is initiated, and to determine reaction velocity.

Acknowledgments

This work was supported by Department of Veterans Affairs Merit Review grant 1I01BX002234 and NIH grant AR1067966 to R.L-B.

References

1. Felson DT, Lawrence RC, Dieppe PA, Hirsch R, Helmick CG, Jordan JM et al (2000) Osteoarthritis: new insights. Part 1: the disease and its risk factors. Ann Intern Med 133:635–646
2. Loeser RF, Goldring SR, Scanzello CR, Goldring MB (2012) Osteoarthritis: a disease of the joint as an organ. Arthritis Rheum 64 (6):1697–1707
3. Finkel T, Holbrook NJ (2000) Oxidants, oxidative stress and the biology of ageing. Nature 408:239–247
4. Blanco FJ, Rego I, Ruiz-Romero C (2011) The role of mitochondria in osteoarthritis. Nat Rev Rheumatol 7(3):161–169
5. Blanco FJ, López-Armada MJ, Maneiro E (2004) Mitochondrial dysfunction in osteoarthritis. Mitochondrion 4(5–6):715–728
6. Wang Y, Zhao X, Lotz M, Terkeltaub R, Liu-Bryan R (2015) Mitochondrial biogenesis is impaired in osteoarthritic chondrocytes but reversible via peroxisome proliferator-activated receptor-γ coactivator 1α. Arthritis Rheumatol 67(8):2141–2153
7. Fernandez-Marcos PJ, Auwerx J (2011) Regulation of PGC-1α, a nodal regulator of mitochondrial biogenesis. Am J Clin Nutr 93 (4):884–890
8. Piantadosi CA, Suliman HB (2006) Mitochondrial transcription factor a induction by redox activation of nuclear respiratory factor 1. J Biol Chem 281(1):324–333
9. Scarpulla RC, Vega RB, Kelly DP (2012) Transcriptional integration of mitochondrial biogenesis. Trends Endocrinol Metab 23 (9):459–466
10. Jornayvaz FR, Shulman GI (2010) Regulation of mitochondrial biogenesis. Essays Biochem 47:69–84
11. Shokolenko I, Venediktova N, Bochkareva A, Wilson GL, Alexeyev MF (2009) Oxidative stress induces degradation of mitochondrial DNA. Nucleic Acids Res 37(8):2539–2548
12. Chang MC, Hung SC, Chen WY, Chen TL, Lee CF, Lee HC, Wang KL, Chiou CC, Wei YH (2005) Accumulation of mitochondrial DNA with 4977-bp deletion in knee cartilage—an association with idiopathic osteoarthritis. Osteoarthr Cartil 13(11):1004–1011
13. Chen L-Y, Wang Y, Terkeltaub R, Liu-Bryan R (2018) Activation of AMPK-SIRT3 signaling is chondroprotective by preserving mitochondrial DNA integrity and function. Osteoarthr Cartil 26(11):1539–1550
14. Yamada K, Healey R, Amiel D, Lotz M, Coutts R (2002) Subchondral bone of the human knee joint in aging and osteoarthritis. Osteoarthr Cartil 10:360–369
15. Blanco FJ, Ochs RL, Schwarz H, Lotz M (1995) Chondrocyte apoptosis induced by nitric oxide. Am J Pathol 146(1):75–85
16. Akasaki Y, Alvarez-Garcia O, Saito M, Caramés B, Iwamoto Y, Lotz MK (2014) FoxO transcription factors support oxidative stress resistance in human chondrocytes. Arthritis Rheumatol 66(12):3349–3358

Chapter 15

Assessing Chondrocyte Status by Immunofluorescence-Mediated Localization of Parkin Relative to Mitochondria

Mohammad Y. Ansari and Tariq M. Haqqi

Abstract

Immunofluorescence staining is a widely used and powerful tool for the visualization and colocalization of two or more proteins and/or cellular organelles. For colocalization studies in fixed cells, one target protein/organelle is immunostained and visualized by one fluorophore and the other target protein/organelle is immunostained and visualized by a different fluorophore whose excitation emission spectra does not overlap with the first fluorophore. Parkin (PARK2) is an E3 ubiquitin ligase which performs ubiquitination of surface proteins of dysfunctional mitochondria to mark them for autolysosomal degradation. Here we describe the immunofluorescence staining of parkin protein and immunofluorescence or dye-based methods to visualize mitochondria and study the colocalization of parkin and mitochondria in primary human or mouse chondrocytes or cell lines.

Key words Chondrocytes, Parkin, Mitochondria, Mitophagy, Autophagy, Mitochondrial dysfunction, PINK1, Immunofluorescence, Confocal microscopy, MitoTracker Deep Red, Osteoarthritis, CCCP, IL-1β, Inflammation, Oxidative stress

1 Introduction

Mitochondrial dysfunction plays a critical role in the process of aging and pathogenesis of several age-related diseases. Mitophagy is a specialized form of autophagy that targets deregulated, dysfunctional mitochondria for lysosomal degradation and functions as a regulator of mitochondrial quality control. Parkin (PARK2) is an E3 ubiquitin ligase that performs the ubiquitination of target proteins/organelles for autolysosomal or proteasomal degradation. Parkin is involved in the clearance of dysfunctional, depolarized mitochondria in chondrocytes, neurons, and other cell types [1, 2]. Loss-of-function mutations in the parkin gene is associated with the occurrence of Parkinson disease [3]. PINK1 (Phosphatase and tensin homolog [PTEN] Induced Kinase 1), a serine/threonine kinase, is produced in the cytosol and is rapidly translocated to

Tariq M. Haqqi and Véronique Lefebvre (eds.), *Chondrocytes: Methods and Protocols*, Methods in Molecular Biology, vol. 2245, https://doi.org/10.1007/978-1-0716-1119-7_15, © Springer Science+Business Media, LLC, part of Springer Nature 2021

the inner mitochondrial membrane where it gets degraded by mitochondrial proteases under normal healthy condition. During mitochondrial dysfunction, the degradation of PINK1 is suppressed and is translocated to the outer mitochondrial membrane where it recruits and activates parkin by phosphorylation of Ser65 [4]. Parkin is predominantly a cytosolic protein under normal condition but it is translocated to dysfunctional mitochondria (Fig. 1) upon the induction of mitochondrial dysfunction [for example by treating chondrocytes with various mitochondrial membrane depolarizing agents such as IL-1β, CCCP (carbonyl cyanide 3-chlorophenylhydrazone), oligomycin, antimycin, FCCP (carbonyl cyanide 4-(trifluoromethoxy)phenylhydrazone), rotenone, and prooxidants such as H_2O_2 and tBHP (tert-butyl hydroperoxide)]. Upon its translocation to mitochondria, parkin ubiquitinates several mitochondrial outer surface proteins such as VDAC1, MFN1, MFN2, TOMM20, and MIRO, and thereby marks the depolarized, dysfunctional mitochondria for clearance by mitophagy [5, 6]. The ubiquitin chains on mitochondrial surface proteins are recognized by p62/SQSTM1, which also binds to LC3 and targets the depolarized mitochondria to autophagosomes [7]. The translocation of parkin is thus a critical event for the clearance of depolarized, dysfunctional mitochondria and thereby for the maintenance of cellular homeostasis. Visualization of the localization of parkin with respect to mitochondria is thus an important assay, described in this chapter, to determine the physiological status of chondrocytes.

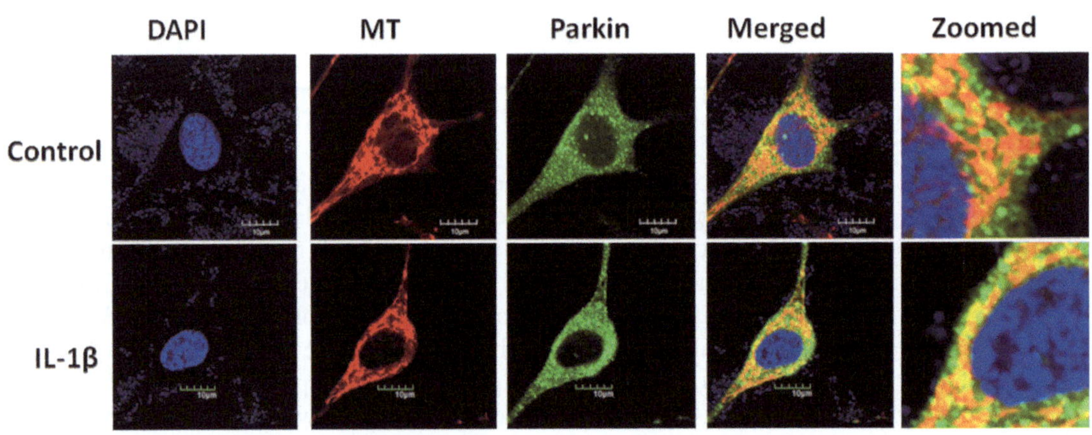

Fig. 1 Colocalization of parkin to mitochondria in primary human OA chondrocytes upon stimulation with IL-1β. Primary human OA chondrocytes were treated with IL-1β for 1 h followed by staining with MitoTracker Deep Red (MT). The chondrocytes were fixed with 4% PFA and immunostained with anti-parkin antibody followed by AF-488 labeled secondary antibody. The nuclei were counterstained with DAPI and images were captured by confocal microscope at 60× magnification

2 Materials

Use deionized water for the preparation of buffers and an analytical grade for all reagents. Follow the guidelines of the institute for proper disposal of hazardous waste.

2.1 Chondrocyte Culture

1. Dulbecco's Modified Eagle Medium (DMEM) with high glucose and L-Glutamine and F12 media supplement (1:1) complemented with 100 IU/ml penicillin, 100 μg/ml streptomycin, and 10% fetal bovine serum (FBS).
2. 8-well chamber slide or cell culture grade round glass coverslips (15 mm, 1.5 thickness) and 24-wells cell culture plate or 100-mm culture dish.
3. Watchmaker's tweezer for handling of glass coverslips.
4. DMEM (with high glucose and L-glutamine and without Phenol red) for staining of mitochondria using MitoTracker.

2.2 Mitochondrial Dysfunction Inducers

1. Cytokines such as interleukin-1β (IL-1β) (final concentration 1–10 ng/ml).
2. Prooxidants such as hydrogen peroxide (H_2O_2) (final concentration 10–100 μM), tert-butyl hydroperoxide (tBHP) (final concentration 10–100 μM), and tertiary-butylhydroquinone (tBHQ) (final concentration 10–100 μM).
3. Mitochondrial electron transport chain uncoupler carbonyl cyanide m-chlorophenylhydrazone (CCCP) (final concentration 10 μM in DMSO), carbonyl cyanide p-(trifluoromethoxy)phenylhydrazone (FCCP) (final concentration 10 μM in DMSO).

2.3 Staining of Mitochondria

1. MitoTracker Deep Red FM (final concentration 100 nM) to stain mitochondria in live chondrocytes (*see* **Note 1**).
2. Anti-succinate dehydrogenase A (SDHA), anti-succinate dehydrogenase B (SDHB), anti-translocase of outer mitochondrial membrane 20 (TOM20), or anti-complex-I, or -II or -III antibody for the immunostaining of mitochondria in fixed chondrocytes (*see* **Notes 8** and **9**).
3. Anhydrous dimethyl sulfoxide (DMSO) as a solvent to dissolve CCCP or FCCP.

2.4 Cell Fixation and Permeabilization

1. Fixative: 4% paraformaldehyde (PFA) made in 1× Phosphate Buffered Saline (PBS), pH 7.4.
2. Wash buffer: 1× PBS, pH 7.4.
3. Permeabilization buffer: 0.2% Triton X-100 (v/v) in 1×-PBS.

2.5 Antibodies, Blocking Agent, and Primary and Secondary Antibody Diluent

1. Primary antibody: anti-parkin antibody and any other antibody listed in Subheading 2.3 (*see* **Note 2**).
2. Control antibody: preimmune serum from the same species (e.g., rabbit, mouse, or goat) as the primary antibody.
3. Secondary antibody: Alexa Fluor 488 (AF-488) or fluorescein isothiocyanate (FITC)-conjugated anti-mouse, -rabbit or -goat antibody (depending on the source of primary anti-parkin antibody) when using MitoTracker Deep Red to visualize mitochondria, or appropriate fluorescently tagged secondary antibody when using antibody for the immunostaining of mitochondria. For example, when using mouse and rabbit primary antibodies, use anti-mouse AF-488 and anti-rabbit AF-594 secondary antibodies, respectively.
4. Wash buffer: 1× PBS.
5. Blocking buffer: 5% (w/v) bovine serum albumin (BSA) or 5% goat serum or 5% FBS in 1× PBS.
6. Antibody dilution buffer: 5% (w/v) BSA in 1× PBS.

2.6 Nuclear Counterstaining and Antifade Mounting media Agent

1. Nuclear counterstain: 300 nM 4′,6-diamidino-2-phenylindole (DAPI) solution in 1× PBS or commercially available DAPI solution.
2. Anti-fade mounting media without DAPI: use commercially available ProLong™ Gold Antifade mounting media without DAPI or make a solution of 2% n-propyl gallate in 90% glycerol.
3. Sealant: nail polish.
4. Fluorescence grade microscope slides.

2.7 Confocal Microscope

We have used 60× oil-immersion lens on an Olympus FV1000 confocal microscope to acquire the images shown here.

2.8 Buffers and Reagents

1. *10× PBS (1000 ml)*: 80 g NaCl (Mr: 58.4), 2 g KCl (Mr: 74.6), 14.4 g Na_2HPO_4 (Mr: 142.0), and 2.40 g KH_2PO_4 (Mr: 136.1). Dissolve these chemicals in 800 ml deionized water and adjust the pH to 7.4 using 1 M NaOH or 1 M HCl. Finally, make up the volume to 1000 ml.
2. *1× PBS (1000 ml)*: Either dilute the 10× PBS to 1× PBS using deionized water or follow the steps below. 137 mM NaCl (8 g, Mr: 58.4), 2.7 mM KCl (200 mg, Mr: 74.6), 10 mM Na_2HPO_4 (1.44 g, Mr: 142.0), and 1.8 mM KH_2PO_4 (0.24 g, Mr: 136.1). Dissolve these chemicals in 800 ml deionized water and adjust the pH to 7.4 using 1 M NaOH or 1 M HCl. Finally, make up the volume to 1000 ml.
3. *Antifade mounting media (100 ml)*: 2 g n-propyl gallate, 90 ml 100% glycerol, and 10 ml 10× PBS. Add 10 ml of 10× PBS to 90 ml glycerol in a beaker and dissolve 2 g of n-propyl gallate

by mixing on a magnetic stirrer. n-propyl gallate is not highly soluble in aqueous solution. Cover the beaker with aluminum foil to protect it from light and stir overnight to completely dissolve n-propyl gallate. Store the solution in aliquots at −20 °C.

4. *DAPI solution*: Make a stock of 10 mM DAPI by dissolving 10 mg DAPI dihydrochloride (Mr: 350.3) in 2.85 ml dimethylformamide. The stock can be kept at −80 °C for long-term storage. Prepare a working stock of 300 nM in 1× PBS, pH 7.4. DAPI is a known mutagen. It must be handled accordingly and disposed of according to the institutional safety guidelines.
5. *4% PFA (100 ml)*: All the procedure should be performed in a ventilated hood. Heat 80 ml 1× PBS, pH 7.4 to 60 °C in a beaker by incubating in a 60 °C water bath. Dissolve 4 g PFA powder in 80 ml preheated PBS. PFA will not dissolve easily and will give a white, turbid solution initially. Add drops of 1 M NaOH slowly to increase the pH to help dissolve the PFA. The solution should then become clear. Check the pH of the solution and if needed adjust it to 7.4 with the help of 1 M HCl or 1 M NaOH. Store the PFA solution at 4 °C.
6. *Poly-L-lysine coating of glass coverslips*: Prepare a 10× stock of 1 mg/ml poly-L-lysine (PLL) in deionized water and sterilize by passing through a 0.22-μm filter. Store the stock at −20 °C. All subsequent procedures are done in aseptic condition in a laminar flow biological cabinet. Before coating, sterilize the glass coverslips with 70% ethanol and air dry. Make a working solution of PLL (100 μg/ml) in 1× PBS, pH 7.4. Put a coverslip into each well of the 24-well plate. Add 0.5 ml PLL solution drop by drop into each well and incubate the plate at 37 °C for 1 h. Aspirate the PLL solution and wash the coverslips with 1 ml of sterile water, 5 times. Finally, rinse the coverslips with 100% ethanol and dry them under UV light for 1 h to overnight. Always use freshly coated glass coverslips for better attachment of chondrocytes.

3 Methods

Carry out all the procedure at room temperature, unless otherwise specified.

3.1 Chondrocyte Culture and Mitochondrial Staining with MitoTracker Deep Red

1. Prepare primary human osteoarthritic (OA) chondrocytes using a standard protocol. Parkin colocalization with mitochondria can also be assessed in normal human or mouse primary chondrocytes. Maintain the chondrocytes in DMEM with high glucose medium, L-glutamine, and 100 IU/ml penicillin/100 μg/ml streptomycin, supplemented with F12 (1:1)

and 10% FBS, at 37 °C and 5% carbon dioxide (CO_2) for not more than 3–4 days.

2. Seed 1×10^5 chondrocytes per well in an 8-well chamber slide in 0.3–0.4 ml DMEM/F12. Give a media change every alternate day. Alternatively, place cell culture grade glass coverslips (*see* **Note 3**) in a 24-well plate and add 1.5×10^5 chondrocytes per well in 0.5 ml DMEM/F12, or put 15–20 coverslips in a 100-mm dish and seed 5×10^6 chondrocytes, or put 4–5 coverslips in a 35-mm dish and seed 1×10^6 chondrocytes. On the day of the experiment, move the coverslips with the help of watchmaker's forceps to the 24-well plate, one coverslip per well containing media with mitochondrial dysfunction inducer. Use no treatment or a solvent alone (such as DMSO in case the compound was dissolved in DMSO) in a control well.
3. Treat the chondrocytes with agents inducing mitochondrial dysfunction. We recommend using 5 ng/ml IL-1β or 100 μM H_2O_2 or 10 μM CCCP for 1 h or using other agents/conditions suitable for the experimental needs.
4. When visualizing the mitochondria using MitoTracker Deep Red, aspirate the DMEM/F12 media and add DMEM (without phenol red media) containing 100 nM MitoTracker Deep Red (*see* **Note 4**) and incubate the chondrocytes at 37 °C in a CO_2 incubator for 30 min.

3.2 Cell Fixation and Permeabilization

1. Aspirate the media and wash the chondrocytes twice with 1 ml 1×PBS to remove unbound dye.
2. Add 0.5 ml 4% PFA to each well of the 24-well plate or 0.3 ml 4% PFA to each well of the 8-well chamber slide. Allow the chondrocytes to fix for 15 min at room temperature *(add PFA in a ventilated hood)*.
3. Aspirate the fixative *(follow the guidelines of the safety office of the university/institute for proper disposal of PFA)* and wash the chondrocytes three times by adding 1 ml 1× PBS per well of the 24-well plate or 0.5 ml of 1× PBS per well of the 8-well chamber slide, each time incubating the plate/slide for 5 min at room temperature. Do not let the chondrocytes dry at any stage throughout the procedure.
4. Add 0.5 ml 1× PBS containing 0.2% Trion X-100 per well of the 24-well plate or 0.3 ml per well of the 8-well chamber slide and incubate at room temperature for 15 min for permeabilization.

3.3 Blocking

1. Aspirate the permeabilization buffer and add 0.5 ml 5% BSA solution in 1× PBS to each well of the 24-well plate or 0.3 ml per well of the 8-well chamber slide. Incubate the plate/slide

for 30 min at room temperature to block nonspecific sites of antibody adsorption. 5% goat serum or FBS in 1×PBS can also be used for blocking (*see* **Note 5**).

3.4 Primary Antibody Staining

1. Prepare control (*see* **Note 6**) and primary antibody dilutions in antibody dilution buffer in a 1.5 ml microcentrifuge tube. For immunofluorescence staining, the antibody dilution range is normally between 1:50 and 1:500. Follow the suggested dilution range from the antibody manufacturer. Vortex the tube to make a homogeneous solution of antibody and spin at 10,000 × *g* for 1 min to collect all the solution (and to get rid of any protein aggregate when using serum for blocking). The minimum volume required to entirely cover the chondrocytes is 200 μl for each well of the 24-well plate and 150 μl for each well of the 8-well chamber slide. The plate/slide can be incubated at room temperature for 1 h, but we recommend incubating the plate/slide at 4 °C overnight in a humidified chamber for proper labeling. Alternatively, to reduce the volume of antibody used when using glass coverslips, put a 100 × 100 mm Parafilm in a 100-mm petri dish and label it in a grid pattern and number them accordingly as in the 24-well plate (normally 10 glass coverslips can be fitted into one 100-mm dish). Place 25-μl drops of appropriate antibody dilution at the labeled spot. Carefully remove the coverslips from the 24-well plate using watchmaker's tweezers and place them inverted on the antibody drops. Make sure the side with the chondrocytes on glass coverslips is down, in contact with the antibody solution.
2. The next day, carefully remove the antibody solution and wash each well with 0.5 ml 1× PBS in each well of the 24-well plate or 0.3 ml 1× PBS in each well of the 8-well chamber slide to remove unbound primary antibody. Wash the chondrocytes three times, 5 min each by rotating at low speed (50 rpm on a circular rotating platform). When using glass coverslips inverted on Parafilm, carefully pick each coverslip, invert it and put it back in the 24-well plate (make sure the cell side is up). Wash each well with 0.5 ml 1× PBS for three times, 5 min each on a low-speed rotating platform.

3.5 Secondary Antibody Staining

1. Prepare fluorophore-conjugated appropriate secondary antibody/antibodies in antibody dilution buffer at the dilution recommended by the manufacturer (normally the dilution range is 1:100 to 1:1000) in a 1.5 ml microcentrifuge tube. Mix by vortexing and centrifuge at 10,000 × *g* for 1 min as in **step 1** under Subheading 3.4.
2. Remove the wash buffer and add 200 μl diluted secondary antibody in each well of the 24-well plate or 150 μl diluted

antibody in each well of the 8-well chamber slide. Protect the plate/slide from direct light by covering it with aluminum foil or by placing it in the dark, and incubate for 1 h at room temperature. Alternatively, when using glass coverslips, the secondary antibody staining can be done in a 100-mm petri dish as in **step 1** under Subheading 3.4 to reduce the volume of antibody used.

3. Wash the chondrocytes with 1× PBS three times as in **step 2** under Subheading 3.4.

3.6 DAPI Staining and Mounting

1. Aspirate the last wash buffer and add a few drops of DAPI solution to cover the chondrocytes entirely. Incubate at room temperature for 2 min.

   Aspirate the DAPI solution and add 0.5 ml 1× PBS per well of the 24-well plate or 0.3 ml per well of the 8-well chamber slide.

2. When using a thin glass 8-well chamber slide, add a few drops of antifade mounting media and acquire the images of the chondrocytes by confocal microscope. When using glass coverslips, label the slides and place one drop of antifade mounting media per coverslip onto the slide. Put a maximum of three glass coverslips per slide. Pick up the glass coverslips carefully from the wells, remove the excess of PBS by touching the edge of the coverslips with a Kimwipes and put the coverslip inverted onto the antifade mounting media drop (make sure the chondrocytes are on the down side of the coverslips, facing the mounting media). Blot the coverslip with a Kimwipes and leave at room temperature for 30 min to dry the excess of mounting media. Seal the edge of the coverslips with a thin circle of nail polish. The slides can be stored in a plastic box at −20 °C for up to 1 year in the dark.

3.7 Imaging

1. Acquire the images on a confocal microscope using a 60× or 100× oil immersion lens (*see* **Notes 7, 10–12**).

4 Notes

1. For mitochondrial colocalization studies, only use a MitoTracker probe that is retained after fixation, such as MitoTracker Deep Red FM # M22426. Not all the MitoTracker probes are retained after fixation. Some mitochondrial probes leave mitochondria upon mitochondrial membrane potential loss. Thus. use the MitoTracker probe whose sequestration to mitochondria is not dependent on mitochondrial membrane potential.

2. When using antibodies to coimmunostain parkin and mitochondria, make sure that the sources of the antibodies are different. For example, when using rabbit antibody for immunostaining of parkin protein, use either a mouse or a goat antibody for the immunostaining of mitochondrial protein.
3. Regular, fluorescence grade glass coverslips can be coated with poly-L-lysine and sterilized in order to use for cell culture.
4. The concentration of MitoTracker Deep Red may need to be optimized in the range of 50–500 nM for best result.
5. When using primary antibody of goat origin, do not use goat serum for blocking or antibody dilution. It will cause high background.
6. Proper antibody controls are the most important part of immunofluorescence staining. Use preimmune serum (rabbit, mouse, or goat depending on the source of primary antibody against parkin or other target protein) as primary antibody control and also include a control for secondary antibody alone to make sure that the secondary antibody is not binding nonspecifically.
7. For colocalization studies, keep the step thickness for scanning as 0.5 μm. Higher thickness may yield false positive results.
8. This procedure specifically describes the immunofluorescence staining of parkin and mitochondria; however, it can be used for the immunofluorescence staining of any other cellular protein.
9. Another way of studying colocalization of parkin and mitochondria is by transfection of an expression plasmid for a GFP-tagged parkin protein and staining of mitochondria with MitoTracker Deep Red. The above reagents can be used to determine the colocalization of parkin with mitochondria in live chondrocytes as well in fixed chondrocytes (after staining with MitoTracker).
10. If there is no fluorescence signal observed under the microscope, optimize the primary antibody dilution.
11. If there is a lot of background signal, optimize the secondary antibody dilution. If further diluting the secondary antibody causes loss of intracellular signal, use 0.01% Triton X-100 in 1× PBS as wash buffer.
12. When using MitoTracker Deep Red to stain mitochondria, optimize the concentration of MitoTracker Deep Red. We found that 100 nM MitoTracker Deep Red is good for visualizing the mitochondrial network in primary OA and normal chondrocytes. If the signal is very weak, increase the concentration of MitoTracker Deep Red. Be careful when using a high concentration of MitoTracker, as it could result in nonspecific staining of other cellular organelles.

Acknowledgments

This work was supported by grants from National Institute of Arthritis and Musculoskeletal and Skin Diseases and National Center for Complementary and Integrative Health of the National Institutes of Health under award numbers R01AR067056, AT007373, and by funds from Northeast Ohio Medical University to TMH.

References

1. Ansari MY, Khan NM, Ahmad I, Haqqi TM (2018) Parkin clearance of dysfunctional mitochondria regulates ROS levels and increases survival of human chondrocytes. Osteoarthr Cartil 26(8):1087–1097. https://doi.org/10.1016/j.joca.2017.07.020
2. Fivenson EM, Lautrup S, Sun N, Scheibye-Knudsen M, Stevnsner T, Nilsen H, Bohr VA, Fang EF (2017) Mitophagy in neurodegeneration and aging. Neurochem Int 109:202–209. https://doi.org/10.1016/j.neuint.2017.02.007
3. van der Merwe C, Jalali Sefid Dashti Z, Christoffels A, Loos B, Bardien S (2015) Evidence for a common biological pathway linking three Parkinson's disease-causing genes: parkin, PINK1 and DJ-1. Eur J Neurosci 41(9):1113–1125. https://doi.org/10.1111/ejn.12872
4. Kazlauskaite A, Kondapalli C, Gourlay R, Campbell DG, Ritorto MS, Hofmann K, Alessi DR, Knebel A, Trost M, Muqit MM (2014) Parkin is activated by PINK1-dependent phosphorylation of ubiquitin at Ser65. Biochem J 460(1):127–139. https://doi.org/10.1042/BJ20140334
5. Ding WX, Yin XM (2012) Mitophagy: mechanisms, pathophysiological roles, and analysis. Biol Chem 393(7):547–564. https://doi.org/10.1515/hsz-2012-0119
6. Youle RJ, Narendra DP (2011) Mechanisms of mitophagy. Nat Rev Mol Cell Biol 12(1):9–14. https://doi.org/10.1038/nrm3028
7. Pankiv S, Clausen TH, Lamark T, Brech A, Bruun JA, Outzen H, Overvatn A, Bjorkoy G, Johansen T (2007) p62/SQSTM1 binds directly to Atg8/LC3 to facilitate degradation of ubiquitinated protein aggregates by autophagy. J Biol Chem 282(33):24131–24145. https://doi.org/10.1074/jbc.M702824200

Index

Tariq M. Haqqi and Véronique Lefebvre (eds.), *Chondrocytes: Methods and Protocols* , Methods in Molecular Biology, vol. 2245, https://doi.org/10.1007/978-1-0716-1119-7, © Springer Science+Business Media, LLC, part of Springer Nature 2021

N

O

P

Q

R

S

MIX
Papier aus verantwortungsvollen Quellen
Paper from responsible sources
FSC® C105338

If you have any concerns about our products,
you can contact us on
ProductSafety@springernature.com

In case Publisher is established outside the EU,
the EU authorized representative is:
Springer Nature Customer Service Center GmbH
Europaplatz 3, 69115 Heidelberg, Germany

Printed by Libri Plureos GmbH
in Hamburg, Germany